IN PRAISE OF THE NEXT DAY CAME TRILOGY

"If you've ever suffered a loss of a loved one, the stories in *The Next Day Came* Trilogy: *Jeffrey: The Injustice of Murder*, *Bud: Homicide Turns a Blue Star Gold*, and *Kathy: Survive Tragic Losses with Limitless Resilience*, will change your life as they changed mine after the loss of my dad and best friend. The emotions of anger, aloneness, and depression became overwhelming, but when I listened to KD and heard her unimaginable life experiences, I felt the love, strength, and empathy she learned through her losses. I found peace in my heart. A must-read!"

—**Private Matt Seliski**, U.S. Army

"This story transcends the world of loss, including the millions of loved ones lost to the pandemic. With raw truth, KD reveals the fragility of life and love. After gun violence took her eighteen-year-old son, unbelievably, two years later, another homicide took her only other child. KD finds the strength to climb from the depths of hell to honor the legacies of her two sons. Bud and Jeffrey changed lives, but KD's stories will change the world. A must-read for everyone!"

—**Debra Vickers**, Lifelong Family Friend

"Courage and leadership come from the heart that split second before the mind has a chance to think. This must-read story, *Bud: Homicide Turns a Mother's Blue Star Gold*, reveals the epitome of these traits in an extraordinary young man, even after he suffered adversity and loss."

—**LCDR Bryan Stern**, United States Navy

"As a Navy mom who has suffered a military loss of a child, this story of another Navy mom whose death of two sons will inspire any Gold Star Mom with a broken heart over the loss of her child."

—**Mona Gunn**, Past National President of the American Gold Star Mothers, Inc. Mother of SMSN Cherone L. Gunn, USS Cole Terrorist Attack 2/14/78–10/12/00

"We will be in the uncomfortable position of what to say to someone who lost a loved one, especially a child, or two children, or all her children as with KD and be afraid to ask: Was this child murdered or lost while serving in the military? How does someone survive this? KD Wagner's unique journey through unbelievable loss, grief, survival, and triumph will inspire people with the strength of humanity. KD Wagner's story allows the blessing of never wondering what to say by giving [us] the strength to honor the child and say their name."

—**Holly Ross,** American Gold Star Mother of LT Rhiannon "PAM" Ross, U.S. Navy 9-11-90 to 10-23-20, Angel to Many. Clear Skies and Fair Winds.

"Kathy's resilience truly shines through in her writing. She takes the reader on a powerful yet vulnerable journey in a way that leaves one heartbroken and inspired at the same time. After reading The Next Day Came Trilogy, there is a deep ache in my heart, but I also laughed and smiled while learning more about her boys."

—**Carly Carruthers**
#1 International Bestselling Author of *Your Epic Book Launch*

"I'm beyond emotionally moved by Kathy's writing. This goes very deep for me as I relate to her in her story big time!"

—**Crystal Davis**
#1 International Bestselling Author of *The Authentic Driven Life*

"What a heart-wrenching, true-life experience no mother (nor anyone else) should have to go through. The loss of a child must be dreadful. To lose what sounds like a mother's soul mate is devastating under any circumstances, but murder? Impossible to imagine.

Her family, friends, doctors, and those closest to her helped her survive the pain of loss, the pursuit to move forward, and her strong desire to help others make it through their journey forward from the loss of someone or something special.

Kathy is an amazing woman. She is truly remarkable with a spirit to share, help, and give."

—**David J. Dunworth**

"Losing a child is unimaginable; losing a child to a senseless murder is unfathomable. To be treated with a total lack of empathy by the local law enforcement is cold and inhumane.

I am in total awe of this woman, as I don't know how I would have made it through, but Kathy found a way to get through with help from her remaining son and her soulmate Kim.

She had me in tears one minute, then smiling the next. From one extreme to the other. It was well written; I felt like she was sitting next to me telling me the story, not like I was reading it. I couldn't put it down."

—**Donna Rauch Lanford**

"Kathy's loss of Jeffrey is heartbreaking. Her story of survival is inspiring. Hard to put down once you start reading. Jeffrey, Kathy, and Bud were the 3 Musketeers. The family photos bring you closer to them. A loving family until the unthinkable happens. A story of love, heartbreak, and determination."

—**Paula Melhorn**

"I caught the author giving an interview, and her story sounded riveting. I wasn't wrong…it was all that and more. It's full of touching and fun tales of her sons, the devastating loss, and the enduring love that brought her back from the deepest grief a mother could endure. It's not only about coping with grief but how to live your best life despite your grief. Highly recommend."

—**Lenita J. Lane**

"Bud lived every day of his life without regrets, striving to succeed, and loving deeply."

—**Marisol Velazquez**

BUD

HOMICIDE TURNS A BLUE STAR GOLD

A True Story

Book Two of *The Next Day Came* Trilogy

Dr. KD Wagner

#1 International Best-Selling Author

© 2023 Kathleen L. Dart Wagner

All rights reserved. No part of this publication may be reproduced, distributed, or transmitted in any form or by any means, including photocopying, recording, or other electronic or mechanical methods, without the prior written permission of the publisher, except in the case of brief quotations embodied in critical reviews and certain other noncommercial uses permitted by copyright law.

Disclaimer: The author recreated events, locales, and conversations from the author's memories. In order to maintain their anonymity and to protect the privacy of individuals in several instances, the names of individuals and places were changed as well as identifying characteristics and details, such as physical properties, occupations, and places of residence. Any resemblance to actual living persons is purely coincidental.

Printed in the United States of America

ISBN: 978-1-7355589-8-1 (ePub)
ISBN: 978-1-7355589-9-8 (Paperback)
ISBN: 978-1-7355589-7-4 (Hardcover)
ISBN: 978-1-7355589-1-5 (Audio Book)
ISBN: 978-8-9867382-0-8 (Activity Book)
Library of Congress Control Number: 2022913221

Published by:
Gold Star Matrix Publishing
690 Main Street Suite #117
Safety Harbor, Florida 34695

Cover Photo: Ariana Studio
Cover Artwork: Ranilo Cabo
Editor: Karen W. Burton

For permission requests, write to the publisher, addressed "Attention: Permissions Coordinator," kd@goldstarmatrix.com

Quantity sales discounts are available on quantity purchases by corporations, associations, and others. For details, contact the author at the address above.

For more information about KD Wagner or to book her for your next event, speaking engagement, podcast, or media interview, please visit LimitlessResilience.com

*To my sons, Bud and Jeffrey, whose lives
and losses made me the person I am today,
and to my love, Kim, who convinced me
to stick around, heal, and tell this story.*

CONTENTS

Author's Note   xi
Acknowledgments   xiii
Introduction   xvii

PART I
LIFE GOES ON

1. Shattered Dream   3
2. College Buddies   7
3. Movin' on Up   9
4. You Have Got to Be Kidding   21
5. The Daily Grind of Waiting   27
6. A Day of Love and Laughter   33
7. War Stories and More   39
8. Phone Phobias   43
9. Nine Phone Calls   47
10. Comedy of Errors (If Only)   55
11. Kim's Recollections   59
12. The Navy Finally Found Me   63
13. Notification Nightmares - Again   67
14. Marisol's Recollections   73
15. Doctor Appointment (Previously Scheduled)   77
16. The Marines Landed   81

PART II
THE BEGINNING OF BUD

17. Life on the Farm   89
18. Surprise Birthday Present   99
19. Bud Joined the Navy   109
20. Navy Showcases   119
21. Bud's 21st / Jeffrey's 18th   129
22. Bud's Recollections - That Dreadful Call   137
23. Brothers Lost in Different Ways   139

PART III
BUD'S NEVER COMING HOME

24. Navy Blues and Déjà Vu	149
25. Funeral Plans—Again	157
26. Family Fiascos	167
27. Coincidences	171
28. Miracles Happen, Or Not!	177
29. The Love Ride (Honoring Bud)	185
30. Payback is a Bitch	187
31. Funerals Suck	193
32. Lynette's Recollections	205
33. Enduring Stories of Love	207
34. TAPS Played for Bud	213
35. Letters and Awards	221
36. The Promise to Keep	227
37. A Day to Honor Bud	233
38. Life Goes On, or Does It?	237
39. Permanent Solutions	245
40. The Promise	251
41. Birthdays Without Bud and Jeffrey	257
42. The Next Day Came —They Continued to Come	261
43. Graduation was In Sight	269
44. The Promise Fulfilled	277
Conclusion	283
Afterword	289
About Gold Star Mothers, Inc.	293
Take the Next Steps	297
About the Author	299

AUTHOR'S NOTE

Thank You for Purchasing: ***BUD: HOMICIDE TURNS A BLUE STAR GOLD***

Life does not always turn out the way we had planned. Through the pain and loss of loved ones, I recognized with clarity that I had two choices: I could end my life or find a new passion and purpose for my new life. My mission became to write the *never-forgotten* legacies of Jeffrey and Bud, my two sons. Their stories allow people to learn that with *Limitless Resilience*, they are not alone, they can survive, and there is hope for a future in a different place.

Listening to my soul, I discovered a sense of peace and learned to use *Limitless Resilience* to move forward. I hope these stories inspire people to find their new purpose and a way to honor their losses. For me, I live every day to make my sons proud and be the person they believe me to be.

Throughout this trilogy, ***The Next Day Came***, I reveal the gut-wrenching truth about loss. You may find common feelings and experiences of loss, but you will also find tools that can help you address the thoughts tumbling through your brain brought on by grief.

Looking back, one of my biggest regrets, when this disaster

happened to me, was that I did not write down everything as it happened in my world. I should have recorded how I felt about the storms in my life. Had I written things down as they happened, I would have a clearer view of how I became who I am today.

As I learned more about coping with my own loss, many recommended journaling and coloring as powerful tools. Portions of this book may trigger intense feelings, good or bad, while reading these stories. If this happens, write those feeling down. To help you with this, I created **The Next Day Came Trilogy Thoughts and Emotions Activity Book**, which provides space to write and color when you need to take the time to process your thoughts. Along with this resource, I am also providing you with the Limitless Resilience Checklist and How to Build Emotional Resilience Video to help you become more resilient in the face of extreme adversity.

These resources can be found here:
www.LimitlessResilienceKit.com
Or scan this QR Code:

Honor Your Losses, Love, and Live Life Limitlessly,

ACKNOWLEDGMENTS

Bud and Jeffrey are the sons I never dreamt of having. I had no desire to produce children of my own, after my childhood. Their love, on both sides of the lifeline, taught me how to live, laugh, and love. They are forever in my heart. As Guardian Angels, they helped me write this book.

Kim Wagner came into my life before the Universe took my sons. Her love, encouragement, and enduring patience allowed me to tell this story and write these books. She became my strength when I had none—the reason I stuck around long enough to heal—and she helped me become the person I was meant to be. I owe her my life and my eternal love.

To anyone who suffered a loss of a child, please know I understand to the depths of my soul the pain and sorrow you endure. There is nothing that compares to the loss of a child; it is the ultimate loss. I honor you in taking the time to read this story and walk with me through the journey.

Truthfully, a loss is a loss. The loss of a loved one, a relationship, a career, or a pet—all hurt. No matter the loss, the process of grief needs walked through to survive. That walk is different for everyone. To heal, one must walk through the pain, find the purpose in that pain, and transform that purpose into a new passion.

I would like to thank the following people for their support and love during my loss, healing, and growth as I moved from grief to the writing of these books: First and foremost, my spouse, Kim Wagner, who supported, encouraged, and became the foundation these books stand upon. Second, my editor, Karen Burton, who made these books a reality. Marisol Velazquez & Joaquin ~ my adopted daughter & grandchild, and Matthew Seliski ~ my adopted son.

I would also like to thank: Dr. Lydie Louis Esq., Leanne VanCamp, Lenita Lane, Nancy & John Wagner, Julie Kessel MD, Pam Lesher, Hope Hoch, Ida Lazar, Sally Jo Hood, Marge Keyes, Julie Arenstein, Richard & Kathy Wagner, Robert & Judy Wagner, Brian & Jane Tattersall, Pat Morrissey, Eleanor Gabardi, Renea Lucci, Deborah Parsons PhD, Julie Humphrey Thometz, Erin Teegarden-Wolbeck PhD, Esther Martinez, Christine Beverly Thomas, Sandy Goss, Amy Luce, Jennifer Schiavone, Kevin Kellin, Jackie Bonafonte, JoAnn Zielinski, Becky Norwood, Natalie McQueen, Heshie Segal, Shannon Gronich, Tami Patzer, Sonya Nagy & Alain Barbera, Dr. Bob & Charlene Levine, Joy Grace Harmony, Keith & Maura Leon, Gary Coxe, Yolanda Mercado, Myrtice Landers, Roberta Brockway, Jean Uffalussy, Toni Gross, Rockie Lynne, Chris Marcelle, Alison M. Wheeler, Orly Amor, and Adam Bricker.

Never say, "That won't happen to me."
Life has a funny way of proving us wrong.
~ Unknown

INTRODUCTION

September 11, 2001, as I turned on the television at 5:45 a.m. in Captain Island, California, I said aloud to Jake and Angel, our poodles, "Is that an airplane stuck into the side of that burning building? Are those bodies falling out of that building, or are people jumping?"

What an odd time of day for an action movie. Suddenly, the picture on the television changed to a plane flying into the side of another tall shiny skyscraper. It looked like New York City to me. I had been there a dozen times.

"Kim, you'd better get down here, now!" I yelled.

She ran down the stairs to the second floor of the townhouse and saw the shocked look on my face.

"Kathy, what's going on?"

"Kim, a couple big-ass planes just flew into a skyscraper. It looks like New York City. This will not be a good thing for Bud!"

"Do you think they flew into those buildings on purpose?"

"Yes, Kim. Why else would anyone fly airplanes into a building?"

"You'd better call Bud and tell him."

Introduction

I dialed Bud's number, he answered, and I could tell he had been asleep.

"Bud, you had better get up. Somebody attacked America. This is not good at all."

"What the hell are you talking about, Mom?"

"Bud, turn on your television. Two planes flew into skyscrapers in New York City."

"I'll call you back. I need to call the Navy and see what I'm supposed to do. Love you, bye."

Instantly, my world shifted again and not for the better.

Kim, my partner of two years, and I had gotten up earlier than usual. She had a flight out of Poison Oak International Airport to Sanford, Oklahoma. She planned to be there for a week, inspecting dozens of apartment buildings for her company based in New York City. The company owned thousands of apartment communities throughout the world.

Kim was already dialing her phone, "I am trying to contact somebody in the New York office."

The phone rang repeatedly; nobody answered. Not sure what we should do, we got in the truck and headed for the airport. As I drove, Kim continued to call her office in New York City to no avail.

"Kim, is there anybody else you can call?"

"Oh, great idea. Let me try Ben in Chicago."

Ben answered, and at least we had communication with Kim's company. She put her phone on speaker so I could hear the conversation.

"Ben, do you know what is going on in New York City? I'm supposed to fly to Oklahoma this morning. In fact, we are headed to the airport now."

"Do not fly anywhere today, Kim. Stay home, stay safe."

We turned around and headed home. By the time we got home, the government had grounded all flights and closed American airspace. We had never expected an attack on American soil. This frightened me profoundly.

September 11th resulted in an immediate change in Bud's military status in the U.S. Navy Reserves. Serving his third year of a

Introduction

four-year commitment, Bud went from a part-time reservist to full-time active-duty.

I believed in my heart of heart that Bud would forever be safe after the murder of his brother, Jeffrey. *They* would never take my other child!

Instead of working one weekend a month, Bud now reported to NAS Point Mugu five days a week. The Navy prepared for a war, but life for most Americans continued the same. Of course, the people who suffered the attack and the loss of a loved one were stuck in the devastation, but the rest of America went back to work or school.

In my experience, this same scenario happens after a funeral. The attendees go home, back to work, and back to their life. Unfortunately, the person who lost a child or loved one never gets over that loss. Left alone, they struggle to find a way to move through it.

This catastrophe yanked me right back into the trauma of Jeffrey's murder. I watched the horrible scenes on the television, hour after hour, day after day, week after week. The world waited for an explanation of the attacks.

I personally felt the pain and the loss of the victims' families. I empathized with those who suffered the ultimate loss of a child. I knew these family's lives would never be the same. I lived with a television on for the next month. The death of thousands of Americans played repeatedly. I watched everything there was about September 11th. It consumed my every waking hour.

Death had consumed my life. Death #1 was my Granddad Adam in August of 2000, Death #2 was Elaina Kim's mom in January of 2001, and Death #3 was my son Jeffrey in March of 2001.

Why was it always one step forward and three steps back in my life?

PART I
LIFE GOES ON

1

SHATTERED DREAM

Unbelievable wreckage hurled into Bud's and my lives with the murder of Jeffrey, my eighteen-year-old son. He had died senselessly during a robbery in Fallen Meadow, Utah, on March 26, 2001. Blindsided, the loss of Jeffrey devastated both of us beyond words. So many things changed, and nothing made sense to us any longer. We could not figure out how Jeffrey could be gone.

For us, there were cumulative effects of Jeffrey's murder. Bud and I struggled with the guilt that we were unable to save him. Together, we shared our anger about the injustice of the justice system, including the police department's nonexistent compassion or empathy. No matter what we did, Jeffrey's murder never made sense to us. Through unbelievable pain, Bud and I did our best to move forward, together.

My partner Kim stuck around with us through this horrendous time. With my history of failed relationships, I certainly was cynical about whether she would stay. Thankfully, she fed us, cared for us, and cheered us on through our worst of times.

Bud and I would not have made it through this disaster without Kim's love and guidance. *Blessed* and *loved* were the only words that described how Kim made Bud and I feel. We believed that Jeffrey watched over us and protected us from heaven. He became our newest Guardian Angel.

Bud and I made a solemn pact to be the people Jeffrey believed us to be. Bud, Jeffrey, and I had grown up together, and we were The *Three Musketeers*. But now, it was only Bud and me, *Two of a Kind*. Lucky for us, Kim showed possibilities of becoming a lifetime member of the group; we would watch and see.

College Adventure

Six months after the senseless loss of Jeffrey, still unable to work from an injury caused by unhooking my truck from the trailer, I enrolled in classes at Surf City College on August 26, 2001. I started college with the utmost hope that my life would find a place in which I could survive.

Dr. Carlson, the psychiatrist I saw for six months, promised me that I would find a *Different Place*. In this place, I would heal and be okay, but I would never be the same. Someday, I could live my life joyfully, with the sense of happiness and love Jeffrey had brought to my life. I strived every day to find that promised Different Place.

The college adventure commenced when Bud reminded me that I spent all his growing-up years telling him and Jeffrey about the importance of an education. Bud remembered The Promise I made to him and his brother.

"Mom, you made The Promise to Jeffrey and me that when we grew up, you would go to college, so you could get an easier job." Bud stood tall, pointed at his chest, and snickered. "Well, we grew up!"

Bud challenged me, and I decided to fulfill The Promise. I always kept my promises to Bud and Jeffrey.

Bud had huge goals, and one of his biggest was his education. He loved learning new things at school. He worked part-time at Mission Community College where he attended classes, carrying a 4.0 grade point average (GPA) and was on the Dean's honors list.

Bud Received His 4.0 Honor Award 2003

Bud was impressed with Kim's knowledge and the places she had reached in her profession. He said, "I want a job like Kim—working at home in my pajamas and going to Hawaii twice a year. That would be the ultimate career. I'm going to get an MBA just like Kim."

"Kathy, you know I dressed every day. I was not in my pajamas," Kim said.

Bud attended college using his GI Bill. The educational benefits through the GI Bill were one of the reasons he joined the Navy at seventeen. As a single parent, I could not afford to pay for his college education. When Bud enlisted in the Navy, he signed an eight-year contract. The first four years were active-duty, which he completed. He was fulfilling his other four-year part of the contract in the Navy Reserves.

Attending college allowed Bud and me to share classes and help each other with homework. We spent time with each other, even if it was doing nothing. Bud and I talked, and the conversations helped us heal.

2

COLLEGE BUDDIES

Despite full-time work in the Navy, nothing deterred Bud from his college education. He was on a mission. We adjusted our schedule and switched to evening and weekend classes at Surf City College. There were times when we did our homework on the phone. If he could not attend classes, I took notes for him, so he could keep up with the class.

Bud and I focused on our studies and strived to do our best. This was particularly important to Bud. Like his mother, he was incredibly competitive and wanted to do and achieve the things his brother could no longer do.

Favorite Classes

One of Bud's favorite classes was a computer class called A+ PC Hardware. The objective of this class was to build a working computer. This was way over my head, as I could barely use a computer at this time. Bud was confident, and we signed up as partners for the project.

Bud and I bought all the parts and assembled the computer. Things got a little stressful when we installed the operating system, and the computer did not work. I called Ralph, a computer nerd friend of mine. He knew the problem instantly and told Bud how to fix it. The computer worked perfectly. We received a letter grade A+ for this class, and Bud had a brand-new computer to use for his schoolwork.

Another class Bud and I enjoyed was Biology. However, Kim was not a happy participant in this course, as we took over her kitchen. One weekend, she came home from a work trip to find a science experiment in progress.

There were ten containers on the kitchen counter filled with various liquids and chicken bones. I do not remember what the ten liquids were. However, I do remember enjoying the chicken wings and conversation with Bud as we placed the bones in the jars.

We recorded how the chicken bones reacted to the liquids in each of the containers. Bud and I found it quite interesting to see what happened; Kim did not. She wanted her kitchen back. We received an A+ in this class too.

Healing Through Learning

Spending this time in college and sharing these learning experiences with Bud was truly a pleasure. We spoke daily, did homework together, and hung out with each other. These activities molded a deeper love between a mother and her son. I cherished this time in college together as it was truly therapeutic and healing.

These classes Bud and I shared allowed us to start healing our broken hearts. We would never get over the loss of Jeffrey, but we made progress and moved forward together. I helped Bud survive and move through the loss of his brother, and Bud helped me survive and move through the loss of my son.

3

MOVIN' ON UP

My time at Surf City College passed quickly. I learned I would graduate in December 2002. Kim loved her job, so we planned to stay in California. I decided to transfer to one of the California state universities. The owner of our townhouse decided to raise our rent when the lease was up, so Kim and I began looking for a home to buy. The only stipulation was it had to be within eighty-five miles of an international airport for Kim's work.

Sticker Shock

We learned quickly that the costs of homes in California were unbelievably expensive. Discouraged, I put a search into Google on the computer for homes under $250,000 in the local area. Homes in the town of a place called Frostbite, California came up, and it was within our parameters of the airport.

We had no idea where that was, but we decided to take a road trip on the weekend to find out. As this was long before Google map or GPS, Kim put the address into MapQuest and printed out a map for us. We headed out on what should have been a two-hour drive.

MapQuest did not account for Poison Oak traffic, and it was horrendous on a Saturday. We ended up at the bottom of a huge mountain range. The map said that Frostbite was up that mountain.

"Do we have to go up there, Kathy?"

"Kim, we drove all the way here; we should at least go look."

"Do you think they have restaurants up there? I am hungry."

There was a sign held by Smokey the Bear: *Welcome to the National Forest.*

I do not know the number of curves and turns there were, but what a climb! The scenery was beautiful. We finally made it to the top of the mountain and found the small town of Frostbite. Our first stop was a small realty office. The people inside were not in the least bit interested in showing us houses, but they gave us a list of homes for sale in the area and a local map.

We went down the street to the first address and began our search. It was a beautiful three-bedroom home on a private drive. From the surrounding deck, there was an unbelievable picturesque view down the mountain to the valley below. The neighbor, out in their yard, told us that on cloudless days, they could see the Pacific Ocean.

Kim and I loved the house at first sight. We went back to the realty office and asked for a tour of the inside. After looking at too many houses, this first house ended up being the home we bought. This would be a big move from the beach at sea level to the mountains at 5500 feet above sea level, with all four seasons. Neither Kim nor I had lived in a four-season area for years. What a change this would be.

View From our Deck, Down the Mountain

The Big School

We drove back down the mountain to the campus of California State University. We found the admissions office, and I filled out an application. They accepted me for the spring quarter. This was a big step in my dream of fulfilling The Promise. I was going to the big school! I could not wait to begin.

"Do you think that things will be okay, Kim?" I asked.

What was I thinking? Had I forgotten what Jeffrey always told me? I hoped there was no curse this time.

Due to Bud's commitment with the Navy and full-time active-duty status, Bud was not able to move up the mountain with us. He decided to stay in the Poison Oak area. He found an apartment, and I helped him move into his new place.

Kim and I spent the weekends up on the mountain in the empty house. We slept on the floor in sleeping bags by the gas fireplace to keep warm. We gave the house a major cleaning and completed all the repairs before moving in, even minor ones. We also had the house painted inside and outside.

"This house smells like old people. Mom, I bet it's the carpet," Bud said.

Next, we installed all new flooring, thanks to Bud's nose. Once our new home was ready, Bud asked his friend Noah, from his active-duty Navy days, to fly out to California. Noah had moved back to Cleveland, Ohio, after he left the Navy.

"Sure Dude, I'd be happy to help move your mom and Kim. I can see Marisol, too."

Bud, Marisol, and Noah had shared a home during his active-duty time in the Navy on the East Coast. When Noah arrived, he noticed the differences in Bud, and the fact that Bud and Marisol were taking a break from each other. This bothered Noah and he talked with me privately.

"I'm concerned about Bud. How is he doing?"

"Bud struggles with his guilt about not saving Jeffrey."

"I know. I can tell that Jeffrey's death is not sitting well with Bud at all; he seems to be in a strange place. It will be exciting to spend time with Bud and Marisol. I miss them. I worry about their relationship, too."

Moving Day

Bud and Noah loaded up our U-Haul and drove the truck up the mountain to our new home in Frostbite on November 26, 2002. Once again, the 26th came into play. Everything in Jeffrey's story repeated on the 26th of a month, and I took these dates as messages from Jeffrey that he was thinking about us.

It was late when we arrived, so we decided to unload a couple of mattresses to sleep on and leave the rest of the truck until morning. We had just settled in when the doorbell rang. Kim answered it to find a lady at the door.

"My name is Natalia. I live across the street. I wanted to welcome the new family to the neighborhood."

She brought us a loaf of freshly baked bread with butter. The bread was still warm and smelled delicious. Natalia stayed and talked for a while.

"I was a famous ballet dancer before I moved here. I danced in the Phantom of the Opera years ago and traveled all over."

Her beauty and personality entranced both Bud and Noah.

The next morning, we unloaded the rest of the truck. Noah got to see the fantastic view from our living room and deck. He was impressed. "Wow, this is so peaceful. I think you and Kim will enjoy living here, Kathy. I hope you can find peace here."

It was Thanksgiving Day, but all the pots, pans, and dishes were still in boxes. There would not be any cooking at our house this Thanksgiving. We found lunch at a local restaurant and ate a turkey dinner.

Bud and Noah took the U-Haul truck back down the mountain and returned it. Kim and I quickly learned that our lives would consist of thousands of trips up and down that mountain. It was comforting to have a place to call our own again. Regrettably, home would never be the same for me, but I wondered if this could be the Different Place that Dr. Carlson talked about. I was not going to get my hopes up about that. It would take time for me to trust anything going right in my life.

Graduation Day

My graduation from Surf City College on December 17, 2002, filled my mind with reflections. Graduating with an Associate of Arts Degree was unexpected for me. Over three hundred students graduated that day. This was an important achievement for me, and it showed Bud that he, too, could reach his dreams.

Milestones Missed

Growing up, I attended a small high school in Pluto, Illinois. There were only one hundred students in my freshman class; it was the largest class to ever attend the high school. During those two years in Pluto, my life changed drastically. My parents finally divorced during my second year of high school. Shockingly, while I was on a summer vacation to Spain, Melinda married a man twenty-five years older than her.

Even more surprising, Melinda sold her house in Pluto, and we moved to Scully, Wisconsin. I spent one-half of my third year in high school there as a Junior. The stepfather owned a house there, which they quickly sold, too. Scully was where I met Artie; less than a year later we married. The math shows that I knew Artie, *in person*, four months before our marriage.

I left Scully after the Christmas holidays, as Melinda bought a home in Bighorn, Utah. This placed me in my third high school in three districts and three separate states in one year. Since there were dissimilar course and attendance requirements in Bighorn, I became a Senior. I finished high school in five months and graduated a year early.

Squandering important milestones was a pattern in my life. A decision, made by Melinda, overruled my desire to attend my high school graduation. Crazy me, I helped Melinda out instead. At seventeen years of age, I skipped graduation and drove Mac, age six, to Pluto, Illinois, for his court-ordered summer visitation with Marine Corps Sergeant Torrance. Today, kids cannot even play outside unsupervised.

After I dropped Mac off, I drove back to Bighorn, Utah—3500 miles round trip. Could this have been where the truck driving bug developed in my soul? I certainly loved to drive, and I had no fear of driving alone across the country.

In August 1977, our new little family went back to Scully, Wisconsin, for the wedding where I married Artie at the age of seventeen. Why did nobody try to stop me from this insanity? Torrance and his new wife attended the wedding and delivered Mac back to Melinda. Divorces are complicated indeed, but I digress.

My First College Degree

When I started at Surf City College, I never thought about a degree, so this graduation ceremony was a bonus to me. My goal was to obtain the necessary credits to transfer to The Big School, California State University. I was on a mission, and that was my focus.

Kim and I had already moved to Frostbite, California, so after

Lynette arrived from Utah, we made the long trip back to Pacific City for the ceremony. Bud and Marisol attended the graduation, along with our neighbor Betty and her dog, Leo. After the ceremony, we went for a celebratory dinner.

Even though I received a diploma, this graduation did not fulfill The Promise I had made to my two sons. My goal was to earn a bachelor's degree that would supply me an easier job. I felt extreme pressure on me to live up to those expectations, but I was determined to keep The Promise I had made to Bud and Jeffrey all those years ago.

Holiday Without Jeffrey

Life for Kim and me calmed down a little as we settled into our new home on the mountain. I waited for school to start again at The Big School. Bud was busy learning new things and earning all the qualifications possible through the Navy, along with studying for his next promotion. Sometimes it amazed me how I had instilled into my children the importance of an education.

"Mom, like you told me: anything I learn, they cannot take away from me," Bud said.

Life on the mountains was different from the hustle and bustle of beach life. The poodles were happy in their pristine environment, with innumerable amounts of squirrels and birds to watch. They were free in the house and enjoyed their walks in the forest.

My medical issues escalated as I continued to struggle with the loss of Jeffrey; however, I kept focused on helping Bud. We talked daily. I tried to be the best mother I could be to him. He needed to know I was there for him and know that he was not alone. I had to be strong for him and help him move forward.

Kim and I decided we would not celebrate Christmas. Neither of us were in the mood. We would have dinner with Bud when he came up the mountain, depending on when the Navy gave him a day off. I had no interest in things that would never be the same again.

New Friends and Neighbors

Nevertheless, Jerry and Geri, our neighbors across the driveway, invited us to their annual Christmas Party. I was unimpressed with the idea of a party, but Kim was like a kid in a candy store. She had something to look forward to. Kim said I would attend, so we were going to the party, even though I had no interest in talking to people or socializing. Kim was the socialite in our family, but she always included me. Still today, Kim continues to manage our friendships and social calendar. I continue to remain skeptical of relationships and making friends, for fear of losing them.

Jerry and Geri turned out to be wonderful people. Painlessly, for me, it was super easy to remember their names. Even more interesting for me was that fact that they both were retired law enforcement officers. In fact, Jerry was a retired lieutenant from two different police departments down the mountain.

The Jerrys wanted to introduce Kim and me to a group of their mountain friends. At the party, we learned everyone there was a member of the mountain's Humane Society. Funny thing, they were all women, except Jerry. We met Peggy, Emilie, Jan, Sue, Bonnie, Tiffany, and dozens more.

When the party started, Jerry sat back in his chair and smiled. He waited for Kim and me, to *get the picture*, as he put it. I figured out right away the situation in the room. It was clear to me that the Jerrys had set this up from the day they met us.

Eventually, Kim became suspicious and walked up next to me at the food buffet. Kim whispered in my ear, "Mr. Jerry is the only man here. Are all these women gay?"

"Kim, if they look like ducks, walk like ducks, talk like ducks—then they must be ducks, right?"

Kim was surprised first, then shocked. "Seriously, Kathy. They're all gay?"

"I'd say the majority of them are, for sure, Kim."

Kim spun away and marched towards Jerry; I followed her. "Jerry, how long have you known that Kathy and I were a couple? Why didn't you say anything?"

Jerry started laughing, and his belly rolled. The look on Kim's face was comical. When Jerry caught his breath, he answered Kim's questions. "Kim, from the day you two moved in at Thanksgiving, we knew and agreed to wait for our annual party. Geri and I decided not to say anything to you about our friends, nor to our friends about you two. Our friends kept asking us if we thought you two were sisters; I told them you looked alike." Laughing, Jerry threw his arms up in the air. He roared, "Welcome to my *harem of women.*"

Jerry loved practical jokes and played both sides of the group. He was in his prime that night, and Kim and I fell in love with him. We were great friends for years. It was good for Kim to have other people to talk to and spend time with as hanging out with me was not always pleasant.

Next thing I knew, Kim talked about playing Bunco. I did not know what that was, but I would not play any games. Kim had a blast at the Christmas party and quickly made new friends. The Humane Society group talked her into becoming a member; I did not join. I was not in the mood to join a group or have conversations with anybody on a regular basis.

Somebody Liked the Damaged Me

As the party went on, a lady name Bonnie introduced herself, and we talked. Well, she asked questions, and I answered them. My story intrigued her even though she seemed to know that I had no desire to talk. The choice I made—to honor my sons by attending college—impressed her. She believed that fulfilling The Promise gave me just enough strength to stay alive and time to heal.

Bonnie was a religious woman and believed that her God needed me to stay on Earth. She understood I had a mission to fulfill and that I should give it my utmost effort. We became best friends, which was odd for me, as I did not make or have friends.

When I had a question about my studies, I had Kim call Bonnie to ask my question, as I did not use the phone. This relay happened dozens of times during our friendship. Bonnie did not judge my strange loathing of the phone. Kim was, and is still, a saint in my life.

Bonnie was an attorney and loved education, so she was always happy to help me. She was willing to meet with me and took the time to help me understand my assignments. No other students in my classes lived up on the mountain, so having Bonnie to help me was a blessing.

Throughout our friendship with Bonnie and her partner Tiffany, we shared dinners in each other's homes and at mountain restaurants. Of the group of women, at the Christmas party that night, these two were the first to pursue a friendship with Kim and me. All four of us were about the same age, creating a strong bond. The people Kim and I usually spent time with were seniors.

A New Family

Kim's stepfather Skip lived in Arizona; she did not see him often. Her mother had died in 2001. My father had disowned me at twenty-nine years of age. The last time I spoke to my mother was the night they told me somebody murdered Jeffrey. Jerry instantly became a father figure to us both. I referred to him as my adopted dad. He was more of a father to me than Torrance had ever been. We shared conversations, meals, and friendship as we became a new family.

The people on the mountain quickly learned my story and were supportive. They showed empathy for my loss of Jeffrey and did their best to make me feel welcome. Patiently, they listened to all my stories about Jeffrey, which I repeated innumerable times.

The Jerrys had three smaller dogs, Megan, Muffy, and Snickers. We had Angel, our seven-pound poodle, inherited from Melinda, and Jake the ninety-pound standard poodle. Angel sat at our sliding glass door every morning and watched for the Jerrys to get up. Her goal was to spend the day with them, sitting on Jerry's lap. We eventually just let them keep her.

A New Addition

Jake was depressed and appeared lonely, so we decided to get a puppy for him. Kim found standard poodle puppies in the *Poison Oak Times* newspaper. We loaded Jake into the car and drove eighty-five miles to pick out a new puppy. We named her Keleli's Midnight Mountaineer, and called her Keleli, in memory of Jeffrey. We previously laid Jeffrey's ashes to rest on a beautiful black sand beach in Maui, Hawaii.

Keleli was a solid black standard poodle with a Russian bloodline. We thought we picked out the most shy and timid girl, but Keleli had a mind of her own and became very independent. She loved the mountains, especially the snow. Jake was happy to have another poodle around the house with him. They quickly became best friends.

Kim and I were industrious, and with the help of our friends, we built the poodles a large wrought iron fenced-in area on the side of the house. We had our repair person, Larry, install two doggie doors —one in the living room and one in the master bedroom. The poodles were able to go in and out to their yard all day long.

Bud Story

When Bud came up to meet the new puppy, Keleli, he decided to take her outside the dog yard to look at the new fence.

I called out, "Bud, put her leash on."

Anyone who knew Bud, knew that he knew everything. He was very much like his mother. *What?*

"Don't worry, Mom. She won't go anywhere."

Kim and I knew differently, but Bud did not listen. Next thing Kim and I knew, we saw the two of them run by the window. Sure enough, Bud was chasing Keleli around the outside of the house. Kim and I laughed at the escapades out the window. Laughter had long before left my repertoire. I rarely laughed in those months and years after Jeffrey's death. Surprisingly, this trivial event made me laugh.

Keleli was fast and enjoyed the escape game; Bud did not seem to find any humor in the game. He finally caught Keleli, but Bud never admitted to us that he should have put her leash on. We never told Bud the fun we had watching him chase Keleli.

However, Kim enthusiastically told this story about Bud for years to come.

4

YOU HAVE GOT TO BE KIDDING

It was hard to believe—*Or was it really?*—but I started college at California State University on March 26, 2003, the second anniversary of Jeffrey's murder. Once again, the 26th continued to be a factor in my life. Hidden meanings were in this number. Were these messages from Jeffrey?

Ironically, that very same week, on March 21, I celebrated ten years of sobriety. In my mind, I easily could have taken that first drink. I seriously thought about drinking every day. Not feeling the pain of the emptiness in the world without Jeffrey would have been wonderful. I truthfully have no idea what kept me from taking that drink. Was Jeffrey himself giving me the strength to continue without the alcohol?

Looking back, I admit I did not do any of the things I knew I was supposed to do as a recovering alcoholic. All I did was not take that first drink. I no longer had a sponsor, so I could not call one. I did not attend meetings; in fact, I had no idea where to find one on the mountain. I was sure they had them, but I never bothered to look. I had no interest in attending a meeting to share the disaster of my life with people who had no idea what I had been through.

The one thing that kept me from taking that first drink was the fact that Bud and Jeffrey were so proud of my sobriety. That, and that alone, was the only reason I stayed sober. I did it for Bud and Jeffrey.

Thankfully, school started again. My career at California State University started with three classes: Expository Writing, Latino Culture, along with Race and Racism. I can honestly say that the Expository Writing class became the ultimate class I took in college. It helped me throughout the rest of my college career, and I recognized that this course was a precursor to sharing this story.

The Big School lived up to its nickname. It was huge and vastly different from Surf City College. Unexpectedly to me, they held my first two classes in a giant auditorium, which held over three hundred students in each class. It overwhelmed me at first, but I quickly learned to sit in the front row, which allowed me to focus on what the professor taught.

Navy News

Three days after I started school on March 29, I received a phone call from Bud. He was still on full-time active-duty at NAS Point Mugu in Oxnard, California.

"What's up, Dude?"

"How's everything going, Mom?"

"Good Bud, I started classes at Cal State this week."

"Cool, do you like it?"

"It is quite different from Surf City College, that's for sure. Two of my classes have over three hundred students in them."

"Those are huge classes, Mom. I have never been in one that big before."

"Yes, it freaked me out at first. The writing class is smaller. I think learning to write will help me."

"I bet it will. You will have hundreds of papers to write in college, for sure."

"I suppose I will, especially as the classes get more advanced."

"Mom, I have news to tell you."

My heart did a flip-flop in my chest. I dreaded bad news. "What news, Bud?"

"My unit is going over to the war zone for Operation Iraqi Freedom."

I knew it. I would not panic, yet. "And what does that mean for you?"

"Well, you are not going to like this, but I volunteered to go first, before somebody with a wife, kids, or a mortgage has to go."

I almost choked out the words: *Are you kidding me?* We knew after 9/11 that the possibility existed, but I hoped this day would never come. I raised an honorable young man. "Bud, I already lost Jeffrey. Do you have to go? Can't you just stay in the States?"

Could I have done anything to prevent Bud from going to Operation Iraqi Freedom? I do not know. He was my only living son, my only living child. However, the Navy was where he wanted to be. How could I not let him be the man he became?

"Mom, I promise you I will come back."

We were big on promises in this family. "Bud, how can you promise me that?"

"I promise, Mom. I can't give you details, but I promise you I will come back."

"Bud, I love you, and I do not want to lose you."

"I love you, too, Mom, and I promise you I will come back. I'll call you later."

"Okay. Love you!"

I hung up the phone. *Bud will be okay, right?* They would not take Bud away from me. He was all I had left. I needed him, and he needed me. I was so tired of pretending to be strong.

Bud and I had made great strides towards that Different Place. We became even closer than we were before. We talked every day, just like Jeffrey and I used to do. Bud convinced me to attend college, and thankfully, I excelled at school. I went to classes, I did my homework, and I stayed alive. The reality of the situation was that I simply existed, but it was the best I could do at the time.

Bud had earned every qualification he could in his military occupational specialty code (MOS) AMH2, an Aviation Mechanic

Hydraulic 2nd Class Petty Officer in the Navy, or E-5, including passing the test for promotion to E-6. Along with his work at the Navy, Bud attended college, and even found time to work part-time at the school as a Veteran's Counselor helping other veterans get their GI Bill payments to pay for their education. He even arranged and put on a Veteran's Day ceremony at his school.

Bud found a best friend while at NAS Point Mugu. Bud and Don spent time together at work and after work. They double dated with their girlfriends. Bud even talked Don into buying a motorcycle so they could ride Harley-Davidson motorcycles together. In return, Don taught Bud how to fish in the Pacific Ocean. Bud was always working on his car or somebody else's car out at the Navy base. He was a busy guy indeed.

Another Navy Call

Later that same week, I received a second call from Bud. This time, he was fuming angry. "Mom, you are not going to believe this. I cannot go to war."

"Wow Bud, great news."

"No Mom, not great news. It's your fault."

"What do I have to do with it?"

"Well, you took all that calcium, didn't you?"

I was really confused at this point. Nothing came to my mind. "What are you talking about?"

"You know, the story you always told me and Jeffrey about how you took all that extra calcium, along with your prenatal vitamins when you were pregnant with both of us. You wanted us to have good teeth."

This made me laugh. "Yes Bud, but what does that have to do with you not going to war?"

"I took my physical today, and they said I cannot go to war because I still have my wisdom teeth."

"What the heck do your teeth have to do with going to war?"

"Well, because you took all that calcium, I never had a cavity. My wisdom teeth did not come through either."

I thought his story was hilarious.

Bud did not laugh. "Mom, this is not funny. It's ruining my life and my career."

"Bud, really, how is this ruining your career?"

"Now I am going to have to stay back two weeks and get my wisdom teeth pulled. The guys will all leave without me, and I will miss the war!"

"I am sure the war will still be going on in two weeks. Bud, I'd rather you never went to war."

"Well, you shouldn't have taken all that calcium."

"Geez Bud, I'm so sorry for being such an unfit mother. I tried to make your life better than mine. I did not want you and your brother to suffer with cavities and going to the dentist all the time. What was I thinking?"

"That's exactly what I am saying!"

"Gosh Bud, how horrible of me to love you and your brother that much."

"Okay, all right Mom, I'm sorry. I know you were just trying to be a good mother."

"Thanks, you and your brother are lucky to have me."

"I know you are a great mom. I love you. Just so you know, I get my wisdom teeth pulled tomorrow. Then I need to take antibiotics for ten days before I can deploy."

"Let me know how everything goes, okay? I will be worried about you, so call me when you can, or have Liana. I'm sorry you will be late for the war."

Liana was Bud's new girlfriend now that he and Marisol were taking a break. Bud had not told me the whole story, but I knew that Bud wanted to get married and have children; Marisol was not ready.

"I'll tell Liana to call you. Sorry I got so upset. I love you, Mom. I didn't mean to upset you."

"It's okay. The whole story was funny, and you made me laugh. I don't get to laugh much anymore, so I enjoyed it. I love you, Bud. I'm so thankful that I have you."

"I love you too, Mom. I'll let you know how it goes."

Bud got his wisdom teeth pulled. Thankfully, everything went well. He was never the best of patients, so I was glad Liana took care of him for the two weeks. He took all his prescribed antibiotics, but he did not take any pain pills they gave him. He was determined to get better and get to that war.

Bud Went to War

Two weeks later, Bud deployed and caught up with his unit. I crossed my fingers and my toes. I focused on the words Bud told me when he promised me that he would come home. There was nothing I could do but hope and pray that he came home safely.

The Next Day Came. Each day continued to come.

5

THE DAILY GRIND OF WAITING

Torrance, a Marine Corps Sergeant in the Korean War, raised his children as *mini-Marines*. This created an extremely dysfunctional childhood for me, along with an irrational sense of obligation. I carried that responsibility with me daily. With Bud off to war, I had to continue with my college education by myself. My thinking was not rational at all; school was all I focused on.

Thankfully, my functioning dysfunctional personality kicked in. The pressure I put upon myself for Bud to be proud of me was immense. I knew in my heart that Jeffrey, up in heaven, was proud of me for attending college. I felt his presence around me all the time. He kept me safe.

Jeffrey always said, "Mom, you are the smartest person I have ever known."

Call from Bud

Not knowing where Bud was or what circumstances he was in increased my stress and anxiety. Time moved slowly; mercifully, we finally heard from Bud. His first stop on his war adventure to

Operation Iraqi Freedom was in Sicily. He called to let us know he arrived safely. "You won't believe it Mom; Sicily is like a third-world country."

Kim, having been to Sicily, Italy, thought Bud's statement was hilarious. "Bud, why do you think Sicily is a third-world country?" Kim asked.

"There are no signs on the buildings. You do not know where anything is. The people all stare at me. I guess because I am so much taller than everybody here. I don't know, but it's not like America."

"Bud, I traveled there. If you get a chance, explore, eat the food, you will love it. Just give it a chance," Kim said.

Of course, everything for Bud was about food. It was valuable for my peace of mind to hear from him and know he was safe. I hoped and prayed that it would not be long until we heard from him again.

For the first time in their lives, I was not with Bud and Jeffrey for their birthdays on July 17. Kim and I went to dinner in their honor and celebrated their birthdays. Kim made reservations for four people, and two people showed up.

The restaurant thought this strange at first, but once we explained that both my sons shared the same birthday and that one son had died and the other one was at war, they were genuinely kind. We received extra special service; the manager had the server bring us two free appetizers before our meal. We ordered our meals and enjoyed stories of Bud and Jeffrey. The meal was unbelievably delicious, and the server even brought us each a slice of birthday cake and ice cream.

School Busy

Back at school, I declared a business major. Bud pursued a business major, so I would too. Who knew? We might even graduate together. I signed up for the second quarter at Cal State. My classes included Accounting I and Science and Technology, both prerequisites for a business degree.

Things were not the same without Bud. I missed being able to talk to him about my classes; we shared everything about school. Even though I had Kim in my life, I felt so alone, in school and the world—lost without Jeffrey, and unable to talk with Bud. I studied hard and completed my classes. My second quarter ended on July 26. Again, the 26th. I received an A in all my classes.

My focus remained on Bud. I tried not to obsess about his safety. He was such a strong person. He was my rock and kept me grounded. The unknown was truly a horrible place for me to be. Bud promised me that he would return, and I tried to believe him. It felt like I was flopping in the breeze, like a leaf in the wind. Loneliness consumed me, even though Kim tried everything she could to fill in the gaps. Bud and I shared an understanding about Jeffrey that even Kim could not. Life was not fair to me at all.

The added stress worsened my medical conditions, and I developed further problems. Fortunately, I was still able to go to the clinic at the university. The doctors took wonderful care of me. They truly felt sorry for me, as a mother who had lost a child to such violent circumstances. I shared my fears of Bud being at war with my doctor. She suggested I discuss these concerns with my partner, so I did.

"Kim, do you believe there could be an evil force in this world that would take Bud from me?"

"No, I don't think so, Kathy. You have suffered enough for one mother. The world just does not work that way. My God does not work that way."

Great News

The best news ever: Bud called and said he was coming home from the war. *Omg!* This was fantastic news. I could not wait to see him. Bud promised me he would come home, and by golly, he was. I was so excited we were going to see him.

Bud and I would talk about school and the things I learned. I had missed being able to talk to him whenever I wanted to. We had become so spoiled, sharing our time, talking, and moving forward.

Mere words could not describe the happiness I felt to know Bud would be safe, back on American soil.

Kim and I were so thankful to have him home again.

Bud Arrived Back from Operation Iraqi Freedom

Here is a photo of Bud and a co-worker back from war. In the background are two of the C-130 planes in their unit.

Help Needed

Previously, Kim and I built a fenced-in yard for the dogs, which allowed the poodles to go in and out of the house through doggie doors. Later, we found out that we built it too close to the road. It was five feet from the road, and it needed to be fifteen feet. So, we had to move the fence.

Kim and I needed help to move the six-foot wrought iron fence that had been set in concrete. We would have to disassemble the wrought iron fence, dig up posts that each had a huge chunk of concrete on the bottom, and knock the concrete off the posts. Then, we needed to dig new holes and reset the posts in new concrete.

It was an immense job for us two ladies. We could easily have paid somebody else to move the fence for us, but it would be much more fun to do the work with Bud. I was glad Bud would be home because we could spend quality time together and share stories. I could not wait to hear about all his adventures. I wanted to know all the details of where he had been and what he had done.

Of course, there would be tasty food involved. It was such a wonderful thrill to be able to talk to Bud again. When he called me, after I told him how happy I was to have him home, I asked him about the fence. "Bud, can you come up to the mountain and help us move a fence?"

"Of course. Why do you have to move the fence?"

"Well, we did not find out the rules when we built it, and now, we know it is too close to the road. So, we have to move it."

"Okay, I'll come help you. I want to see you while I'm home."

In my excitement to see him, I did not catch *while I'm home*. "Awesomeness, I can't wait. Could you talk a couple of Navy guys into coming with you? We have to dig up fence posts with cement on the end of them."

"Just what I wanted to do, Mom. At least, we can spend time together."

"That was exactly what I was thinking, Bud."

"I'll see what I can do."

"Promise them home-cooked food. What would you like?"

"Mom, you know I want beef roll-ups."

"Okay, I'll let Kim know."

"I have lots of stories to tell, and I have a lot of pictures to show you."

"I can't wait. I've missed you so much."

"I missed you too, Mom. I love you."

"I love you too, Dude! You promised me you would come home, and you did! See you this weekend."

To say I was thrilled would understate the emotion I felt about Bud being home. It would be so fantastic to see him. I missed him so much. We had so many things to talk about. It was a dream come true for me to have Bud home safely.

The fear of wondering whether Bud would make it home from

the war was unbelievable. I would not even talk to Kim about it or give it voice in the Universe. Every damn time a phone rang, my heart did flip-flops. I was so afraid it would be bad news. My brain circled in a constant state of flight-or-fight, a psychological mess.

What a blessing to know that Bud was back in America, on American soil, and *safe!* Kim decided we should have a party. There was no better reason in my mind for a party than to celebrate Bud, safely at home. We invited all our new friends over to share in our joy of Bud's return from war.

It was amazing. Something in my life went right for once: Bud was home.

6

A DAY OF LOVE AND LAUGHTER

Bud and Don came up incredibly early on Sunday August 24, 2003. Kim and Bud got busy in the kitchen and prepared the beef roll-ups. They always enjoyed cooking together. It took a little time to assemble the beef roll-ups, but they were so worth the effort. While Kim and Bud prepared the food, Don and I enjoyed a cup of coffee and teased Bud about his cooking skills. It was not long before Kim and Bud had the beef roll-ups simmering in the skillet; we could smell the delicious dinner to come.

We decided we had better get busy on the project of moving that fence. I had already bought all the supplies we needed for the day's project. Bud, Don, Kim, and I all used wrenches to take the wrought iron panels off the posts. With that done, Kim and I dug the old posts out of the ground. Bud and Don took sledgehammers and broke the cement blob off the bottom of the posts. The four of us worked well as a team, and the work went smoothly.

We laughed and talked, as Bud and Don shared stories of war and traveling. I had school stories, and Kim shared work and vacation stories. After three hours, we decided to stop for lunch. We were dirty and sweaty, so we sat out on the deck while we ate. Our discussions continued to flow as we ate. We were deep in conversation when Bud abruptly stood up, turned, and looked directly at me, as he changed the subject.

"Mom, I want you to know that if I die on my motorcycle, I died happy."

My heart skipped a beat or two. *No way!* I was not going down that road again. I lost Jeffrey, and I barely made it through that. It was enough to endure Bud going off to war. I would not consider the fact that I could lose Bud, period. "Bud, we are not going to talk about this. I would not make it through another loss!"

That was the end of the conversation. We never mentioned it again. I would not even think about the words that Bud had just put into the world. I put that conversation right out of my head.

Important Facts

Bud owned a 2001 Harley-Davidson FLRT Road Glide motorcycle that he bought after Jeffrey died. He rode it everywhere and absolutely loved it. Bud was an incredibly careful rider, and I felt comfortable with his riding abilities.

Before he got his motorcycle license, I made him take the motorcycle safety course in Arizona where we lived. I not only made sure he took it, but that he passed it. We took it together, and we both passed. He received his motorcycle license at the age of sixteen.

At each of his duty stations in the Navy, they required Bud to take a safety course. He wore the proper attire, including leather boots, gloves, safety driving glasses, and a helmet. When he rode his motorcycle on base, he wore a reflective vest. He was a safe and professional motorcycle driver.

Bud and His Harley-Davidson Motorcycle

After lunch, we went back to work and finished the fence. We cleaned up and headed into the house. People started to show up for the party.

Those beef roll-ups smelled delicious after cooking for hours. We used my mother's recipe. Melinda always served them with homemade mashed potatoes covered in the gravy made with the sauce left in the skillet—comfort food to the max.

Beef roll-ups are simple to make and delicious to eat. We substituted cream of chicken soup for the mushroom soup because Bud hated mushrooms.

"If there was one mushroom in the recipe, Bud would find it!" Kim said.

BEEF ROLL-UPS

Prep Time: 45 minutes
Cook Time: 5 hours
Serves: 6–8 people

Ingredients:

- 3 pounds thinly sliced round or sirloin steak (have butcher tenderize)
- 1 package Pepperidge Farm Stuffing (blue bag)
- 3 regular size cans cream of mushroom (or cream of chicken) soup
- Water, empty can x 3
- 1 large container sour cream

A large electric skillet with lid if available, or stovetop skillet with cover
Toothpicks (preferably round)

Directions:

1. Prepare the stuffing per package directions.
2. Pound steak flat with a tenderizing hammer.
3. Cut out any pieces of fat on the steak. Place the fat in the skillet, melt it to grease the skillet, then dispose of it.
4. Cut the steak into approximately 6-inch x 3-inch strips. Place a spoonful of stuffing at the end of the strip of meat and roll it up to the end. Place 3 toothpicks into the meat, through the stuffing, and out the other side of the meat to hold them together. Repeat for all the meat and stuffing.

5. Place the meat into large electric skillet and brown on all sides.
6. Mix the 3 cans of cream of mushroom soup with 3 cans of water, then pour over browned meat.
7. Cover and simmer on low until the meat cooks completely.
8. When done, remove *only* the meat from the skillet, place in a large bowl, and cover with aluminum foil.
9. Remove any loose toothpicks from the remaining sauce in the skillet.
10. Add container of sour cream and stir in skillet. Heat to a boil, making gravy.
11. Serve the meat with the toothpicks still intact and pour gravy over them.
12. Remove toothpicks while eating.
13. Do not forget the mashed potatoes.

7

WAR STORIES AND MORE

We shared an excellent dinner with Bud and our friends. It was so reassuring to have Bud home. It was a wonderful, enjoyable day. We forgot about all the manual labor of moving the fence. The boys shared their stories of being in the war and how different the world was over there.

Bud and Don told a story about their climb up Mount Etna in flip-flops. Why they did not have shoes on, I have no idea.

"The opportunity arose, so we took it," Bud said.

"Bud, you are a dork!"

Bud was always an adventurous guy. They talked about riding in the net seats on the long trip to Sicily aboard a C-130 cargo airplane. Bud and I laughed; we had that same adventure when he was a kid. We flew from Hawaii to California and back in an Air Force plane.

The friendship that Bud and Don shared was good to see. When he was growing up, we moved so many times due to my work, that he did not have long-term friendships. The Navy turned out to be an excellent experience for Bud; he grew into a strong young man.

I know Bud went to Sicily, Bahrain, and Iraq, as he shared photos with us at our dinner. However, I do not know the exact areas Bud went to during the war, as he could not tell us. Bud brought home a bottle of wine for Kim from Sicily. He changed his original story about Sicily being a third-world country after the other places he went to over there.

Bud shared two stories about Bahrain. The first was a photo from a lavish dinner with friends. He said the food was great, and the after-dinner cigars were better.

Bud Smoking a Big Fat Cigar!

The second story made everyone laugh.

"The Navy told me to be inconspicuous while I was in Bahrain. I asked them how I would go about looking inconspicuous as a six-foot-three-inch tall, light-skinned male with white-blonde hair?"

"What did they say?" Kim asked.

"They laughed and told me to do my best."

"And how did you do?" I asked.

"Mom, there was nothing inconspicuous about me being there. I stuck out like a sore thumb."

Everybody laughed and enjoyed the stories Bud shared. It was fun to have them there to share our day. I had missed Bud so much. What a blessing it was to just spend time with him and listen to him talk. I was enormously proud of him.

After dinner, Bud played with the poodles. Jake and Keleli loved his attention. He also spent time with Jerry. Since Bud did not have a relationship with his father, he loved and respected Jerry's opinions. The feeling was mutual. Jerry did not have a good relationship with his children or grandchildren, so Bud was like a grandson to him.

Dessert Time

The evening was getting late, so Kim decided it was time to enjoy her famous Dutch apple pie for dessert. We each had a piece with a scoop of vanilla ice cream on it, exactly the way Bud liked it. He loved Kim's homemade pies.

Everyone shared laughter, joy, and stories. Kim and I were so thankful Bud and Don came to help us move the fence and shared this time with us. We were sad to see him go back down the mountain, but they had other things to do.

Shocking Revelation

Just before they left, Bud made another announcement. "Mom, I will be headed back to Sicily again, back to Operation Iraqi Freedom."

I about fell to the ground. "Are you kidding me? You just got back. Why do you have to go again?"

There were no adequate answers. Bud and I said our goodbyes. We always said *I love you* and shared a hug and kiss goodbye.

After Bud and Don left, our friends told me what a wonderful young man Bud was and how lucky I was to have him for my son. He had laughed and had fun the entire day with us. It was wonderful to share Bud with all our new friends. I was so thankful to see Bud again and talk with him. It had been a wonderful day indeed. I planned to cherish it in my heart and mind until I saw him again. I could not believe how I missed Bud while he was gone. It was so hard not to be able to talk to him every day and share all my school stories. I missed hearing about the Navy and all his adventures, but what I missed the most was having someone to talk about Jeffrey with.

Bud would be gone again, and I would wait patiently for every email or phone call from him. This going to war thing did not make me happy. Bud had gone before and returned, so I was confident that he would come back again. Either way, I really had no choice in the matter. I was just his mother, who loved him with all my heart.

Bud did promise me, didn't he? I do not remember hearing those words when he left. I must have missed it when he said, "Mom, I promise I'll be back."

8

PHONE PHOBIAS

A not so funny story: Kim realized that I would not, or could not, answer the phone. Between Kim and me, we had four different phones, all of which could be ringing, and I would not answer them. Kim ran like a wild animal to answer whichever phone rang, even if that phone were sitting right next to me. My not answering the phone drove Kim crazy.

If I knew the call was from Kim, 99 percent of the time, I answered. After the announcement of the murder of Jeffrey on my phone, I refused to use the phone. A habit that still lasts today, Kim takes and makes phone calls for me. She has memorized all my vital statistics and pretends to be me on the phone. Occasionally, they catch her in this game, but she has it down to a science.

If people asked me a question, I answered, but in all honesty, I had no interest in conversation with people, certainly not on the phone. There was nothing but *bad news* there. When Kim realized that the phone was a continuing problem, she felt I should talk to Dr. Carlson about the situation.

Visiting Psychiatry

Since I refused to see a different psychiatrist, she called Dr. Carlson and made an appointment for August 26 at 11 a.m. Yes, another blessed 26th. I drove the three-hour trip down to Otter Beach, California through Poison Oak traffic and arrived a half hour early. I had not seen Dr. Carlson in over two years. The last time I saw her was before we moved to the mountains.

It was like old home week for us, as we truly had become more friends than doctor-patient. We talked about Jeffrey. I told her about school and how well my courses were going. We discussed the Navy calling Bud up for active reserves when 9/11 happened. I shared that he had gone to Operation Iraqi Freedom, returned, and was leaving on his second tour.

I explained to her that Kim figured out I would not, or could not, answer the phone. Dr. Carlson said, "It's obvious to me that you are not answering the phone because you're afraid that you'll get horrible news again. It makes perfect sense. Kim told you on the phone that somebody murdered Jeffrey. Therefore, you equate the phone with bad news."

"I have no interest in talking to anyone at all, much less on a phone."

"That makes sense, Kathy. If you do not talk to people, you cannot receive bad news. It truly doesn't matter whether it's on the phone or not, does it?"

"No, it does not. Kim practically kills herself to get to the phone. Remember the O.J. Simpson commercial that used to be on television, where O.J. ran through the airport to catch a flight, hopping over suitcases and people to get to the gate on time?"

"Yes, I remember that commercial well."

"Kim hops over me, the dogs, or whatever, to get to that phone before it stops ringing. The phone can ring until hell freezes over as far as I'm concerned."

"Kathy, Bud going to war may have triggered a reaction in your brain intensified all the grief and anger over Jeffrey's loss. Just know that it will continue to take time."

"Okay, I will tell Kim that you said I was okay. She will laugh at that one, for sure."

"You don't have to answer the phone. Let the machine answer, then listen to the message. You can call them back. Get a new cell phone that has caller ID."

"I'll tell Kim I need a new cell phone, too."

"Have you started writing your book yet?"

"Nope!"

"Please do. I am glad you came to see me. Call me anytime you need to."

"Thank you, I appreciate it. I'll let you know when I write the book."

I drove home and gave Kim the good news.

9

NINE PHONE CALLS

March 28, 2001, was the date I found out that somebody murdered Jeffrey, even though the actual murder took place on March 26. I found myself wondering how it could even be possible that exactly two years and five months, to the day, of *Next Days* had come to pass. Life had a way of changing whether I wanted something different or not.

Before that dreadful day in 2001, my life was looking good. I had a fantastic relationship with Kim. My oldest son Bud had just moved in with us after finishing his active-duty in the U.S. Navy, and I was waiting to hear from Jeffrey, who wanted to move home with us too. My thoughts drifted to the four of us, dreaming of a happy, normal, loving family.

I remembered peacefully driving my tractor-trailer down a freeway in California, listening to an audiobook, when Kim called. With no other choice, Kim told me on the phone—as I drove at 65 miles per hour—that somebody murdered Jeffrey! My life changed that very instant and has never been the same. How somebody could just shoot Jeffrey remained incomprehensible to me. That dreadful night played over and over in my mind.

On the morning of August 28, 2003, when the house phone

rang, I took Dr. Carlson's advice and did not answer the phone. I did not answer the second time it rang, nor the third. Since Kim was not home, she did not run to answer these calls either. Kim had left two hours before to help our friend Emilie rake leaves somewhere on the mountain.

I decided to check and see if anyone had left a message. The red light was flashing, so there was a message. I thought it might be Kim, since Jake had not felt well before she left. I pushed the button. Strangely, the call was from Liana, Bud's new girlfriend.

"Kathy, please call me as soon as you can; it's important."

Liana sounded upset. While I contemplated what to do, my cell phone rang. It was from the same number on the answering machine, so I knew it was from Liana.

Oddly, I felt jolted by a lightning strike. Déjà vu flashed through my brain, reliving the night Bud's father called telling Bud about his brother Jeffrey's murder. Bud revealed to me later, *"Mom, I wish I had never taken that call."*

Phone Call #1 at 9:00 a.m.

"Liana, what's going on?" I asked.

"Kathy, I am sorry to bother you, but I cannot find Bud. He left here this morning for the base. He said he would call me when he got there, and he hasn't."

She was talking a hundred miles a minute. I could tell Liana was in tears. "Okay, let's not panic. There is surely an explanation," I said.

"Kathy, it's on the news. Motorcycle accident. U.S. 501 freeway closed. Not Bud?"

"Whoa, Liana! There are hundreds of accidents on the freeways in Poison Oak every day."

"But he has not called me!"

"Okay Liana, I will make some calls, and I'll call you back."

For God's sake, there is not a f**ked-up force in the world that could be that mean. Is there? They already took Jeffrey, what the freaking else do They want from me?

BUD

Phone Call #2 at 9:15 a.m.

I called Bud, no answer. *Okay, he is busy.* I decided to call the Navy. *Why not?* I was his mother. *The Navy must talk to his mother, right?* The number was in my cell phone, as Bud had given it to me two years before when he mustered into NAS Point Mugu.

Phone Call #3 at 9:20 a.m.

I called the U.S. Navy. Somebody answered, I do not remember who. They identified themselves professionally with military unit, rank, and name.

"May I speak to Bud, please?"

The phone went dead in my hand.

Are you kidding me? They hung up! What the hell?

Phone Call #4 at 9:25 a.m.

I called the Navy again. Nobody answered.

Phone Call #5 at 9:27 a.m.

I called the Navy one more time. Nobody answered.

Rationalization worked its way into my mind. *Okay, the Navy is busy getting ready to leave for the war in Iraq, right? They do not have time to talk to a mother calling her son.*

I wanted answers, and I knew there was more than one way to skin a cat, as the saying went. Trained as a law enforcement officer years ago, my dysfunctional, responsible personality kicked into high gear. My brain clicked into investigation mode.

I turned on the television and found the local news. I waited for the traffic report and got an update on accidents in Poison Oak on the freeways. Sure enough, there was a motorcycle accident on U.S. 501 freeway, westbound. Unfortunately, there were no photos or other information.

That did not help me, but it gave me another idea. They

mentioned the California Highway Patrol officer spokesperson. At this point, *things were still okay in my mind.*

Looking back, I should have called Kim to come home and help me.

Phone Call #6 at 9:40 a.m.

I called the California Highway Patrol office nearest the accident reported on the television. I doubted they would answer my questions or give me any names, but I would pull the ex-law enforcement card and see if I could get lucky. An officer answered the phone and identified himself by division, rank, and name; I identified myself. I told him my story and concerns.

The officer confirmed that there had been a motorcycle accident on U.S. 501 freeway westbound. Regrettably, he could not give me the name of the victim in the accident, but he told me where they took the victim. "The victim was taken by ambulance to St. Mary's Hospital."

"Thank you for the information."

I hung up. *Now what should I do?* Obstinate, I would not give up. This had become a mission, and I was obsessed. It was like an unknown force in the Universe drove me to continue. My heart and head pounded, my blood pressure must have been sky high, but I could not, and would not, stop.

Looking back, once again, why the hell did I not call Kim? She could have taken over and called all these people. Illogically, there was no stopping me; my fingers dialed before I could change my mind.

Phone Call #7 at 9:55 a.m.

I googled, then dialed the number for St. Mary's Hospital. A lady answered the phone. I identified myself and spat out the whole story to her. I told her that my son Bud rode a motorcycle, would have been on U.S. 501 freeway, was in the Navy, and was going back to Operation Iraqi Freedom.

Finally, I took a breath. *Why the hell is Kim not the person talking on this phone?* Confusion was settling into my brain, but through the fog, I kept pushing forward. Possessed, I hunted the answers that I pursued. "I need to know the identity of the person involved in the motorcycle accident this morning."

"I'm sorry ma'am. I don't have that information. I'll transfer you to the emergency room."

I could not speak, so I said nothing.

Phone Call #8 at 10:05 a.m.

"Emergency room charge nurse," the voice said.

The title sounded official. I repeated the whole story again. She might have thought me crazy, felt sorry for me, or she was a mother and understood my emotional state of mind, either way, she started asking questions.

"What is your son's full legal name?"

I told her Bud's full given name; a name I never called him.

"When is his birthday?"

I told her.

"Does your son have any tattoos?"

"Yes," I said, and gave a description of one on his hand.

At 10:10 a.m. on August 28, 2003, my life, as I knew it, ended.

"I am sorry to inform you the young man is your son," she said.

I said nothing.

"I am even sorrier to tell you that your son did not make it," she said sadly.

"But you don't understand." I whispered to nobody, as I felt like somebody hit me in the chest with a sledgehammer. There was nothing left in me; all I could say were those four words. *But you don't understand.*

What kind of evil f**ked-up world do I live in? There is no way They could take Bud, my only other child. How dare They? Who in the hell do They think They are? This is so freaking wrong.

"Ma'am, we did everything we could," she said.

"But you don't understand."

"I'm sorry, but I don't understand what?" she asked.

"This *cannot* happen again!"

"What cannot happen again?"

"You just don't understand." I collapsed to the porch. *How did I get out onto the porch?* My brain could not put this information together. Jeffrey was dead, so Bud could not be dead. The nightmare of Jeffrey's murder slammed into my heart like a freight train—full force into my chest. This made no sense. *How can Bud be dead?* They could not both be dead.

The charge nurse must have been concerned for my sanity or sensed that I was in shock. "Ma'am, is there anybody there with you?" she asked.

"No."

"Is there any way you can get somebody to be there with you?"

"My neighbor. But you don't understand."

"Ma'am, do you have another phone? Can you call your neighbor?"

"Yes."

"Okay, stay on the line with me. What is their name?"

"The Jerrys."

"Okay, call them, please."

Phone Call #9 at 10:20 a.m.

I called the Jerrys on the house phone, and Geri answered.

"Can you come over here?" I asked.

"Why?" Geri asked.

"She wants you to come over here."

"Kathy, who wants me to come over there?"

"A charge nurse, phone."

"I don't understand. I'll come right over, hang on."

Why am I holding two phones, one in each ear? I hate phones. Where is Kim?

"Ma'am, are they coming?" the charge nurse asked.

I about jumped out of my skin; I had forgotten about the charge nurse. "Yes."

"Ma'am, what is your name?" she asked.

"Kathy."

"Kathy, I am so sorry to be the one to tell you about your son."

"But you don't understand."

"Kathy, what is going on? Are you okay?" Geri asked, reaching the stairs, out of breath.

Geri had a breathing condition called chronic obstructive pulmonary disease (COPD) and was on oxygen twenty-four hours a day. It was hard for her to climb up the stairs. When she got to the top of the stairs, she tried to catch her breath.

I was also trying just to breathe any air into my lungs.

"Is your neighbor there now?" the charge nurse asked.

"Yes."

"Okay, I'm going to let you go now. Kathy, I am so deeply sorry for your loss. I wish I did understand, so I could help you. My thoughts and prayers are with you and your family."

She was gone, just like that. I wanted to say: *What family?*

"Kathy, what is going on?" Geri asked.

"Bud is dead."

"Kathy, I know Jeffrey is dead. What is wrong? Where is Kim?"

"BUD IS DEAD!" I screamed.

10

COMEDY OF ERRORS (IF ONLY)

The *Soul Train* roared into town; the gates crashed down over the crossing, while the red lights flashed on the gates. *All Aboard!* The *Soul Train's* whistle blew as Death #4, Bud, plowed me down as I stood on the tracks.

Observers on the outside might have seen this day as a comedy of errors. Unfortunately for us, it was a horrific, catastrophic chain of events. Incredibly, once again, somebody on the phone told me my child was dead. I hated the freaking phone. My mind swirled in confusion. *How can this have happened twice?*

They call this kind of things déjà vu, the Twilight Zone, or another realm of the Universe, right?

Really, *They* did not have to hit me over the head *twice* for me to understand this loss of a child thing. I got it the first time with the force of a sledgehammer to my heart. Truthfully, it had sucked the life out of me then; I certainly did not want to replay that tune. Time stopped; reality escaped me. There was no air to breathe, but I must have been breathing, as I was still alive. My heart raced so fast that it should have exploded through my chest.

Can my heart just burst and end my misery, please, pretty please, with sugar on top?

"BUD IS DEAD!" I screamed again.

The expression of utter annihilation on Geri's face verified that she finally grasped what I said. The more I tried to process the information, the more it was impossible to accept. Frozen in time, Geri and I knew this horrible, unbelievable, atrocious secret. *What should we do with this secret? If we do not tell; the secret is not true, right?*

"Kathy, who told you Bud was dead? How did you even find out?"

"Charge nurse."

"How were you on the phone with a charge nurse, Kathy?"

I held my head to keep it from exploding; the pain was unbearable. "I—don't—know!"

"What charge nurse? What hospital?" Geri asked.

What planet am I on? "I—don't—remember."

"Kathy, I think you better sit down," Geri said, as she tried to help me.

"DO NOT TOUCH ME!" I screamed.

"Kathy, where is Kim? We need her home."

"Emilie—rake leaves."

"Does she have her phone?"

"Yes."

"Kathy, have you tried to call her?"

"No."

"Why not?"

"I— don't—know!"

"Let me get Jerry over here first," Geri said.

"But you don't understand."

Geri Phone Call #1 at 10:45 a.m.

I never wanted to see, dial, talk on, or for that matter hear a freaking phone ring ever again.

Geri demanded, "Jerry, come over to the girls' house now!"

BUD

Geri Phone Call #2 at 10:50 a.m.

"Kim, you need to come home; we've had an emergency," Geri said. Unfortunately, cell service on the mountain was horrible. "Kim, can you hear me?" Geri put her phone on speaker, as if that would help. "Kim, Bud is dead."

"Geri, tell Kathy to take him to the vet."

Obviously, Kim thought the call was about Jake, the standard poodle who did not feel well.

"No, Kim. You need to come home right now."

"I can't understand. Take him to the vet. Emilie and I will meet her there."

"Kim, come home now. BUD IS DEAD!" Geri screamed.

"Emilie and I will come to the house. You're breaking up."

Hopefully, Kim could tell something was wrong—terribly wrong.

"Kim should be here with all of us. Oh my God, how could this have happened?" Geri asked aloud.

Jerry Arrived

Jerry walked over to the stairs, along with Joe and Grace, our other neighbors.

"Geri, what is going on? Where is Kim?" Jerry asked.

"Bud is dead," I said.

"Yes, Kathy, we know Jeffrey is dead," Jerry said.

"No, Jerry. Bud is dead," Geri said.

I could tell from the look on his face Jerry realized what she said. He appeared stunned to his core. I thought: *Not another dead person today. People, I cannot take anything else.*

"Are you kidding me? Geri, what the hell happened?"

Jerry kicked instantly into police mode and started grilling for answers. "How does she know about Bud? Has the Navy been here? When did Bud die? Where did it happen? How did Bud die?" I said nothing.

"Jerry, I have no idea. Kathy was on the phone with a charge nurse when I got here. She just repeats the words, *but you don't understand,*" Geri said.

"Where is Kim? Why isn't she here?" Jerry asked.

"I don't know, Jerry. I tried to call her, but reception was terrible. She is with Emilie on the mountain somewhere. She is coming home now, I think."

"Oh my God. How much more is this girl supposed to deal with?" Joe asked.

"I can't believe this," Grace said.

Just then, Kim and Emilie roared around the corner and skidded into our private street. Emilie was driving like a maniac. The truck skidded into our driveway, Kim and Emilie jumped out.

Can lightning just strike down and kill me so I can go to heaven with my two sons, NOW?

11

KIM'S RECOLLECTIONS

With a rare day off, I decided to spend time with our friend Emilie. She needed help with her yard work business. It would be peaceful to be outdoors and share conversations with Emilie instead of being stuck in my office. Emilie picked me up at 7 a.m., and we went over to a house across the mountain top to rake leaves.

Geri Phone Call #2 at 10:50 a.m.

Things were going along well until Geri called. I could not understand what she said. Cell phone reception on the mountain was terrible, at best. I thought for sure that there was something wrong with Jake, our standard poodle. Before I left home, he appeared to not feel well. That was in the back of my mind while I raked. With the phone call from Geri, I felt something bad had happened to Jake.

All I heard on the phone were garbled noises. "Geri, have Kathy take Jake to the vet, and we'll meet her there."

Geri yelled something, which was strange, as Geri never yelled. Since I could not understand what she said, I motioned to Emilie.

"We need to go back to the house; something happened."

Emilie unhooked the trailer full of leaves from her truck. We got in, and Emilie tore out of the yard. We headed for the house as quickly as we could on the curvy mountain roads. Emilie's driving was faster than I would ever drive; she was like a wild woman. We raced across the top of the mountain.

When we turned down our private road, I saw a crowd of people standing in our driveway. Jerry, Joe, Grace, Geri, and Kathy. I did not see any of the dogs. I looked at Emilie in a panic. "This is not good; I don't see any dogs at all."

"Yeah, not good at all," Emilie said.

I thought Jake was dead. We got out of the truck and hurried over to the group. "Is Jake, okay?" I asked.

Nobody said anything. Everybody stared at Emilie and me.

"Yes Kim, Jake is fine. This is way worse," Geri said.

Immediately, I looked at Kathy and realized something was wrong, very wrong! Kathy did not say anything; she looked like death warmed over. She didn't seem to realize I was there. She was crying, talking to herself, and disoriented.

"Bud is dead," Geri finally said.

"How can you say that? Are you kidding? We just saw him. Bud was here on Sunday. You had dinner with him, remember?"

"Kathy called me. I really don't know what happened," Geri said.

"Kathy called you?" I asked.

"Yes, she said they wanted me to come over here. I asked her who, and she said, *the charge nurse*."

"What charge nurse? How was Kathy talking to a charge nurse?" I asked.

I was not sure Kathy even understood the conversation Geri and I were sharing. She was not responding to the conversation; she just stood in the same spot, lost, and confused.

"Exactly, Kim! I knew something was seriously wrong. I came right over and found her on the deck up there. She hung up the other phone and told me that Bud is dead."

"But how does Kathy know that Bud is dead?" I asked.

"I thought she was talking about Jeffrey. She just says, *but you don't understand*. I'm not even sure she is here with us right now."

Emilie, grasping what Geri said, went pale as a ghost.

"Emilie, don't die on us now. We need to figure this out," I said.

"Oh my God, Kim, I'll call Peggy, she'll know what to do. Do you think we ought to call an ambulance? Kathy doesn't look good at all," Emilie said.

Peggy was a retired nurse. We all went to her for medical or veterinarian diagnoses, depending on the patient, human or pet.

"Emilie, get Peggy here ASAP," I said.

I looked at Kathy. Emilie was correct; she looked horrible. For that matter, none of the seniors in the driveway looked good either. I had no idea what to do or how to help Kathy.

How can this happen again? Kathy looked so devastated. She had just started to come around a little. Her relationship with Bud was going so well. She enjoyed school and was on a mission to honor her sons. *How can these terrible things continue to happen to her?*

"Let me see what I can find out from Kathy. We can wait until Peggy gets here before we call the ambulance, but get her over here now," I said.

I walked over to Kathy and reached out to her, but she backed away in shock.

"DO—NOT—TOUCH—ME," Kathy screamed.

"Kathy, please, can you tell me what happened?" I asked.

Through sobs and tears, Kathy tried to convey the events. She rambled through the story from the phone call with Liana to the call to Geri. The narrative was not clear at all. I did my best to put the story together. Whether we called an ambulance or not, I had my doubts as to whether Kathy would make it. I was not sure of her mental state at all.

The story was mindboggling. How had Kathy made phone calls when she never answered or talked on the phone? She should have called me. This baffled me how Kathy found out all this information.

"I never called Liana back. BUD IS DEAD," Kathy yelled.

"Okay Kathy, calm down. I'll call Liana; it will be okay. Just stay calm, and I'll tell her. We'll figure this out, Kathy, I promise," I cried.

Kathy slid down onto the ground, held her head, and bawled. "I can't—not again—but you don't understand. Bud cannot be dead!"

12

THE NAVY FINALLY FOUND ME

My world turned upside down, destroyed again. People stood in my driveway, taking turns asking me questions. I could see their lips moving, but I could not hear what they were saying, as I had nothing more to tell them. *How can They take Bud away from me?* He was all I had left of my life. Sitting on the ground, my head spun and pounded. Bud had been at this very house on Sunday. The Universe could *Just Go to Hell*.

I have no recollection or memory of Kim calling Liana. I had no idea what Kim said or what happened to Liana; there was nothing after the realization that I never called her back. I sat in the driveway and cried. Lost in my own world of sorrow, I isolated inside my own head.

Kim told me later that Peggy showed up; I have no memory of this happening. Collectively, the group in the driveway decided that since I already had a 4:30 p.m. doctor appointment, it was best to leave it up to the doctor to decide my fate. With Peggy's blessings, Kim would drive me to the clinic, go in with me, and explain to the doctor what happened.

Kim later divulged, deeply shocked by the senseless loss of a young man, that nobody was thinking clearly. Irrationally, they decided to take turns watching me until it was time to go to the doctors. I was sitting on the ground, rocking, and mumbling, *but you don't understand*. "Kathy, had I truly known what was going on in your head, I assure you I would have called 911 and had them take you by ambulance to the hospital. At the minimum, you should have been committed for a seventy-two-hour evaluation, but hindsight is 20/20."

Shockingly, at 12 o'clock p.m., my cell phone started to ring up on the deck, sitting where I had left it. It would be an ice-cold freezing day in hell before I ever answered that freaking phone again. Kim, on the other hand, jumped over me and raced up the stairs to answer the phone before it stopped ringing. She looked at the caller ID number.

"Kathy, it's the U.S. Navy."

With no acknowledgment from me, Kim answered the phone. Kim ran down the stairs, put the phone on speaker, and held the phone out to me for a response.

"Kathy, tell the Navy that I can speak for you."

I was not going to speak to the Navy, or anybody else, if I had my way.

"Kim can speak," I said.

With that, the Navy told Kim what they wanted to say. I doubt they realized that Kim left the phone on speaker, nor that eight people were standing, or sitting in my case, in my driveway. Our home had excellent cell service, so everyone heard clearly what the Navy had to say.

"We know Kathy knows that her son Bud is dead," the Navy said.

We know. She knows. Who knows? Déjà vu exploded in my brain. I instantaneously flashed back to the night I learned of Jeffrey's murder and how the police department bungled the notification by calling Bud and Jeffrey's father, Artie. He in turn called Bud and ruined his life, leaving him to tell me about the murder of my son, his brother. Thankfully, Kim made that call for Bud.

The Navy phone call was repeating the same bungled notification processes. This was the definition of insanity, wasn't it?

Looking back, interestingly, the Navy never said how they knew that I knew Bud was dead.

"Casualty notification officers went to Kathy's house; they had the wrong address," the Navy said. The Navy obviously had my phone number, so why had they not called me? Déjà vu confirmed. Another notification fiasco! *Should I just lie down and bang my head on the concrete of the driveway?* If the Universe had the balls to take both my children, could they not have the heart to just tell me?

"Bud did not update his mother's move to Frostbite in November," the Navy said.

Blame the dead victim, really? How convenient can it get? There is no doubt in my mind that Bud updated the Navy of both his and my moves. Bud was always compliant with the U.S. Navy. What difference did it make? My children were both dead!

Kim, always much nicer than I, let it go; she gave the Navy the correct address.

"Casualty notification officers from the Marine Corps base will be there by 4:00 p.m.," the Navy said.

"Kathy has a doctor's appointment down the mountain at 4:30 p.m.," Kim said.

"They will wait for her return," the Navy said and hung up.

What will they do if I never came back? How long will they wait? I did not want to talk to the Navy, or anyone. They could all go to hell as far as I was concerned. I wanted to go to heaven to be with my sons.

We found out later the casualty notification officers waited for my return at a restaurant in downtown Frostbite—that very same restaurant where Kim and I shared Thanksgiving dinner with Bud and Noah in 2001. These military personnel being in Frostbite caused quite a scare in the small town. People knew exactly what these two officers were there for: a lost loved one, but whose?

Authors note: The military sends two officers to the home of a loved one who died while serving on active-duty in military service. The two officers show up in a dark sedan, knock on the door, and proceed to ruin the unsuspecting families' lives with the notification of death. This surely is an improvement to the telegram sent to the family in earlier years.

13

NOTIFICATION NIGHTMARES - AGAIN

Shocked and bewildered, Kim eventually convinced me to move upstairs and into the house. What did it matter where I was? Bud was dead. Kim wanted me to sit down and rest, but I could not, so I walked aimlessly around the house. I could not find a tiny little piece of stable ground to stand on, so I kept moving. Nothing made any sense to me.

"Kathy, do you want me to call people and tell them what has happened?"

"Whatever—not calling—ever."

"Okay, I'll start with your family."

Notifications Suck

At 1:00 p.m., Kim started to make those dreaded phone calls. She felt people needed to know that my only other child, Bud, was dead. *Really, what horrible thing have I ever done that was so bad that They had to take both my children?*

Kim's first call was to Lynette in Utah. That was déjà vu for sure. I went through that horror two years before with a phone call to tell her about Jeffrey. This was wrong, so very wrong. Suddenly, a thought popped into my head. I yelled down the hall. "Kim, tell her to sit down."

Things that burst into my brain surprised and baffled me as to why I even remembered, but it upset Lynette when I called and blurted out that Jeffrey was dead with no warning. Later, she jokingly said, "Kathy, if this ever happens again, make sure you tell me to sit down first, okay?"

Kim put the call on speaker. "Lynette, this is Kim. Are you sitting down?"

"Why?"

"You made Kathy promise to tell you to sit down first before she told you any bad news."

"I'm sitting. What's happened now?"

"Bud has been killed."

"Are you freaking serious?"

"Sadly, Bud was killed this morning."

"How?" Lynette asked.

"He was on his motorcycle headed to the base, an accident or something. I don't have any details."

"This is freaking unbelievable."

"Bud's girlfriend called Kathy this morning and told her there was a motorcycle down on the freeway and asked Kathy to call the Navy as Bud's mother."

"Kathy answered the phone?" Lynette asked.

"Yes, I wasn't home. Kathy is not clear at all—something about calling the Navy, the Highway Patrol, a hospital. A charge nurse told her Bud was dead."

"What is she, the freaking FBI? That is crazy, Kim. Seriously, how is she doing?"

"Not good. She has a doctor's appointment at 4:30 p.m."

"Kim, make sure you go in with her."

"I will. No way I would let her drive—she would probably drive off a cliff at a high rate of speed."

"Are you sure she shouldn't go to the emergency room?" Lynette asked.

"I truthfully don't know. Right now, she just walks around the house saying, *but you don't understand.* She goes out the door in the living room, across the deck, and back in the master bedroom, then down the hall back to the living room, and out again."

"Will she make it through this?"

"I don't know. After the doctors, the Marines are coming."

"Wait, what for? Kathy already knows."

"I guess they have to officially notify Kathy that Bud is dead."

"Geez, I don't think she needs to hear that again, do you?"

"No, but maybe they will have more information."

"Kim, I'll head up there today."

"Okay, I'll let Kathy know. Thanks, Lynette."

"I hope she makes it through this."

"Lynette, I honestly don't know. She barely made it through Jeffrey. Can you tell your kids?"

"Sure, I'll tell them they lost another cousin. This has got to stop!"

"Lynette, she doesn't have any more kids to lose. Kathy needs all of us right now."

"I don't know how she is even breathing. I would not be standing upright, and I'd be heavily sedated," Lynette said. "Kim, where is Kathy's gun?"

"Oh my God, I hadn't thought about that. I will sneak it over to Jerry's for now. Thanks, Lynette."

More Notifications

Next, Kim called Wyatt in Plateau, Washington.

"Wyatt, it's Kim, I'm calling about Kathy."

"What happened?"

"Bud was killed this morning."

"Was he over in Iraq still?"

"No, Bud came home this month. There was a motorcycle accident or something. We don't have clear details."

"How did she find out?"

"I'm not sure at this point. She really is not making sense, Wyatt."

"Understood. Has the Navy been there yet?"

"The Marines are on the way, but I have to take Kathy to the doctor first."

"Why the Marines?" Wyatt asked.

"They are closer to our house, I guess."

"Okay, call me when you get more information, please."

"Will you call your parents and let them know? Kathy has not spoken to your mother since Jeffrey's murder, and she does not speak to your father at all. Would you contact Mac as well?"

"Yes, no problem."

"I do not know if Kathy will make it through this, again," Kim said.

"Tell her I am sorry and let me know what you find out from the Marines. Thanks."

Kim's Family Notifications

Kim called her stepfather Skip in Arizona.

"Kim, is there anything I can do?"

"Right now, Skip, I have no idea what to do for Kathy. I'm taking her to the doctor. I'll let you know when we know more."

"Okay Kim, tell her I love her, and I am so very sorry this happened again to her."

Kim also called her three brothers. When Jeffrey died, they really did not know me, but they had learned my story and would want to know what happened. First Kim called River and Kelly who lived in Arizona. Kim had completely forgotten that River was traveling and coming up the mountain to our house over the weekend.

River, shocked by the news, talked with Kim for a while, which helped her. Kim was trying to be strong for me. "Kim, let Kathy know how deeply sorry we are. If there is anything we can do, please let us know."

Next, she called Ben and Joyce in Wisconsin. Joyce could not believe the horror of the situation and the devastation that I must be feeling. "Kim, please tell Kathy how sorry we are for her loss."

Lastly, Kim called Jack and Nadia in Washington. They were heartbroken with the news. "Let Kathy know that we are thinking of her, and she is in our prayers."

Final Artie Notification

Kim called Artie, Bud's father.

"Artie, it's Kim, Bud was killed this morning. I do not have any details yet, but I wanted you to know."

Kim did not remember Artie's reaction, only that he said, "We will be at the funeral."

Why Artie wanted to attend another funeral for a child he never really cared about—and had thrown out of his home twice before—was beyond me. Artie wasted his opportunities to ever know his two sons; why show up at their funerals?

Fortunately, or not, depending on how you looked at my life, Artie and I had produced no other children. Therefore, there would be no other notifications, funerals, or contact with Artie and Gail after Bud's funeral. How could I even think those words? My life was over.

There were so many calls to make. Kim dialed and shared the horrible news. Since we did not have any details of the funeral service, all these people would need callbacks when Kim obtained more information. Nobody could believe that I had lost my only other child—me least of all.

14

MARISOL'S RECOLLECTIONS

Bud and I were taking a break from our relationship. We were still best friends, but we decided we needed space. Bud wanted to get married, but I was not ready yet. I knew I would forever be a part of his family. I was there when Jeffrey died; that bond we shared would never go away.

Marisol and Bud

I had recently separated when I met Bud and Noah in Maryland. I told Bud I just needed time before I made that kind of big decision again. He was not happy about my choice, so he moved out and started dating other people. It was not what I wanted, but I understood his decision. We talked every day on the phone. It was strange, but we shared every part of our lives. We talked about work, school, friends, and even the other women.

August 28, 2003, 8:00 a.m., as I got ready for work, I vaguely overheard a snippet of the morning news as it played in the background.

"Traffic is piling up on the U.S. 501 freeway westbound. There's a motorcyclist down."

Something sparked in my mind. *Please, not you. No, it cannot be you. What can the odds be?* I shook it off. Thinking nothing more about the morning news, I went to work for the day. When I received the call from Kim later that morning, I knew instantly that something was wrong, very wrong. Kim had never called me at work. I answered the call with absolute fear and dread.

"Marisol, I'm sorry to tell you that Bud is gone."

I was sure I had misunderstood what Kim said. "What did you say?"

"Bud is gone; he was killed. I am so sorry, Marisol."

"Kim, you have got to be kidding, right?"

"Regrettably, no. I'm so sorry. I don't know any details other than Bud was on his way to the Navy base on his motorcycle."

"Oh my God, Kim!"

"It is unreal. Kathy is not good at all."

"I don't know how she is even alive."

"Me either, Marisol. I'll call you when we know more."

We hung up the phone. *Bud cannot be DEAD!* I heard a blood-curdling scream. Later, my coworkers told me it was me who screamed. I told my boss, "I have to go home; I have a family emergency."

I walked out of work and got into my car, but I do not know how I got home. I do not remember driving, stopping at traffic

BUD

lights, or anything at all. Somehow, I got to my apartment—the same apartment Bud and I had shared.

The next few days were a blur. I waited at home, staring at the phone. I knew Bud would call because he called every day, but the phone did not ring. I kept telling myself: *There must be a mistake. There must be a mistake.*

Abruptly, I thought about Kathy. My heart broke for her and for myself. I was devastated. I could not imagine how Kathy's heart could beat any longer. Surely, her heart had exploded into pieces. First, she lost Jeffrey, and now Bud; it was all so inexplicable.

How can the world be so cruel to one woman? How can anyone lose two sons in two years and survive?

15

DOCTOR APPOINTMENT (PREVIOUSLY SCHEDULED)

August 28, 2003, 4:00 p.m., Kim drove down the mountain to California State University and took me to the clinic for my 4:30 appointment. The clinic did not normally treat students for long-term illnesses, but the medical staff knew the story of my loss of Jeffrey, so they continued to treat me.

So far, they had treated me for diabetes, high blood pressure, high cholesterol, and migraines.

Upon arrival, they assessed my blood sugar. It was extremely low at 51. They gave me a carton of juice to raise my blood sugar quickly. Before Liana's call, I had eaten breakfast and taken my diabetic medication. After I learned of Bud's death, I had not eaten or drank anything.

Medical Advice

"I am surprised you have not gone into a diabetic coma," the doctor said.

That might be a good thing. I would not have to think about this rotten life anymore. Why can I not just fall over dead, so I can join Bud and Jeffrey in heaven?

Since I did not want them to *Baker Act* me or put me in the psychiatric hospital against my will, I kept quiet. If they had known the thoughts in my head, things might have turned out differently at this visit. Kim explained what happened.

"Oh my God, this is horrible. I am deeply sorry, Kathy. You have suffered so much already. Nurse, get a valium to calm her down."

The doctor needs a valium too, I thought to myself. The senseless death of Bud had shaken her up.

"Kathy, it is extremely important that you eat something every three hours."

"Not eating."

"You must eat something. Eat a cracker or a piece of cheese."

Kim paid more attention to the doctor than I did. She must have taken notes or something because she still uses those exact words with me even today.

"It is extremely important for you to track your blood sugar every single day."

"Hate blood sugar."

"Too bad, do it anyway," the doctor said.

"Okay, fine."

"Kathy, do you feel suicidal?" the doctor asked.

"Not at this moment," I lied.

"Well, if you start to feel like that, go to the emergency room right away, or come in and talk to me at any time. Didn't you see a psychiatrist when your other son was killed?"

"Yes."

"She saw one for six months before we moved up to the mountain," Kim said.

"You should think about talking with her again, or somebody. This has been another horrible event in your life and talking to somebody would surely help."

"I'll give Dr. Carlson a call and let her know what happened to Bud and Kathy's status. She met Bud along with Kathy two or three times, when Jeffrey died. I will get her recommendation. Thanks for reminding me."

The doctor was a feisty lady but extremely concerned and caring. I do not know if I would have made it through without her and the doctors at California State University.

Non-Medical Advice

The most memorable part of the doctor visit was at the end of the appointment. "You girls listen to me. Do not let the military or the funeral home take advantage of you. You can buy a casket for a lower price at Costco," the doctor said.

When Kim heard the doctor say this, she about fell to the floor but managed to keep a straight face. The doctor must have had a horrific experience with a funeral home when a loved one died.

"You girls check it out. I am not fooling; Costco will deliver it right to the funeral home for you."

"Okay doc, we'll do that. Thanks for the advice," Kim said.

Kim thought this was hilarious and never forgot it; she tells this story still today.

"Be careful. People take advantage of those who are grieving," the doctor said.

I know the doctor was concerned for me as a student who was paying for my education. We thanked her for her help and left the office. I was only fooling myself with my mental health. As I look back, the doctor or medical staff should have realized the dangerous state of mind I was in, but who knows? It really was not their field. I should have told the truth about what I was thinking and feeling. The thing that stopped me was the fact they would lock me in a hospital. So, I survived as I always had, alone, inside myself.

16

THE MARINES LANDED

Back up the mountain, the Casualty Officers from the nearby Marine Corps base must have waited at the end of the street, watching the house for Kim and my return. We no more than got home when the two officers rang the front doorbell. They were both in Marine dress uniforms, which made me cry.

Raised by Marine Corps Sergeant Torrance and seeing those uniforms brought forth a deep-seated fear in me, in addition to the real reason they were there. Mercifully, these Marines were both genuinely kind. The male Marine Corps officer was a chaplain, and the female Marine Corps officer was a nutritionist. They were compassionate and concerned, especially after they learned I lost my son Jeffrey two years earlier.

Of great interest to Kim and me, they shared what they knew about Bud's death. "According to the report, a witness was in a car behind Bud on the U.S. 501 freeway. They were in the exit lane for the base when a car came across four lanes of the freeway and hit the motorcycle. The vehicle slowed; there were three people in the car, two of which turned to look back. Then, the car took off at a high rate of speed. There was no doubt with the witness that the people in the car knew they hit the motorcycle."

"Did they catch the people?" Kim asked.

"Not yet. One of the witnesses chased down the fleeing vehicle, recorded the license plate number, and gave it to the California Highway Patrol. According to the California Highway Patrol, because the people did not stop at the scene, the accident became a felony hit-and-run—causing death, or a homicide," the male Marine Corps officer replied.

Looking back, neither Kim nor I understood what the Marine Corps officer said. We had no idea where he got this information. Clarity would come later.

The officers explained what would happen in the next few days. They said Navy people would come to the house called casualty assistance calls officer (CACO). These people officially represent the Secretary of the Navy. They would help me through the rest of the process and explain more about the funeral and what would happen next.

The female Marine Corps officer had a degree in nutrition and gave suggestions to Kim about my eating. As we learned from my doctor earlier at the University, I had to eat something regularly. I could not go without eating all day. In one day, two professional people shared this same information with Kim. There was no getting away from it. Geez! We thanked them for all they did, and they left.

After the officers left, my head was about to explode. I just wanted to lie down and die. I had no desire to be on this Earth. Sadly, things did not get any better. Next thing I knew, the damn phone was ringing again. Kim received another unwelcome phone call for me.

Organ and Tissue Donations

Tragically, I was the one who had to take the call, so she handed me my phone. This call was from the Center for Recovery and Education (CORE). They apologized for calling at such a horrible time, but they had verified that Bud wanted to donate his organs and tissues.

If I could have said one simple *Yes* to all their questions *just one time*, it would have been easier, but I had to answer every single one of the unbelievably detailed, horrible questions individually. There must have been a hundred questions about the donation of organs, tissues, and parts for transplant.

This process was horrendous, but I understood that it was a gift to the people who needed life-saving transplants. I knew Bud felt strongly about this issue, and his kindness would help innumerable people. Surprisingly, this gift would pop up in my life years later, but that will be another story.

This, once again, was an incredibly horrible day in my life. In fact, it was the second worse day of my life—the first being the day I learned somebody murdered Jeffrey. The amount of pain a human being could endure was preposterous.

There was no rational thought in my mind. Instead, questions raced through my head. *How can it even be possible that both my children are dead? What am I supposed to do? How will I survive?*

Exhaustion set in. I was tired of trying to push through things. Life was simply too hard to deal with. I wanted to go to sleep and never wake up again. *Is that too much to ask for?*

Shock and Anger

"Kim, I am going to lay down for a while."

Kim came and sat on the side of the bed by me. "Kathy, I am so incredibly sorry. When Geri called, it frustrated me that I could not understand her. I was so sure something happened to Jake. It never even entered my mind that something happened to Bud. When I found out that Bud was dead, it threw me for a loop. I honestly believed that after Jeffrey died so violently, Bud would be safe."

Raised Catholic, Kim believed in her God. I said nothing.

"Kathy, I am a little lost myself. I believed God would never take Bud from you. I thought my God did not work that way. This is not sitting well with me at all. I am so deeply sorry."

"I know, Kim. I thought Bud would be safe. What kind of cruel world do we live in?"

"I just do not know, Kathy. I can't believe you called so many people."

"It was crazy. A mysterious force took over me. Something in my gut, or heart, made me keep dialing the phone. It said: *Do not stop until you get the answer.* I had to prove Liana wrong."

Kim sighed, "But you made all those calls only to find out that Bud was dead. That was the horrible part Kathy, and the whole disaster of it all—Bud is dead."

Beginnings and Endings

Lynette arrived from Utah after the Navy CACO Officers left; she did not come alone. She brought Nette, one of her twin daughters. Lexie, the other twin daughter was on her way from Wilmington, California. Both Lexie and Nette brought their children with them.

Life and death were ironic. When Bud was born, Lynette came to my home with her twin daughters, Lexie and Nette, both five months of age. Flash forward twenty-four years—Bud died, and Lynette came to my home with her twin daughters, Lexie and Nette. People say coincidences are not real. I disagree! Innumerable strange intricacies happened in the lives of my two sons.

All my hopes and dreams were gone. Bud and I had worked so hard to move forward and try to heal. We wanted to be the people Jeffrey believed us to be. We had learned together how to survive and live in a world without Jeffrey. How the hell was I supposed to go on without my two sons? The *Three Musketeers* existed no longer. *Two of a Kind* was gone. Left alone, *One* was not the place I wanted to be.

For twenty-five years, Bud and Jeffrey were my sole focus and purpose. Everything I did, all the long hours, all the crappy jobs, all the harassment I tolerated were to give Bud and Jeffrey a better, happier life than I grew up with. I worked so hard to teach them to be good men and raise them into adults.

What do I have to live for? My sons are both gone. Will I ever find another purpose for my life?

I had no answers.

The Next Day Came. The freaking Next Day would not stop coming.

ACTION STEP
MY LIFE AS I KNEW IT ENDED

As I learned more about coping with my own loss, many recommended journaling and coloring as powerful tools. Portions of this book may trigger intense feelings, good or bad, while reading these stories. If this happens, write those feeling down. To help you with this, I created **The Next Day Came Trilogy Thoughts and Emotions Activity Book**, which provides space to write and color when you need to take the time to process your thoughts. Along with this resource, I am also providing you with the Limitless Resilience Checklist and How to Build Emotional Resilience Video to help you become more resilient in the face of extreme adversity.

These resources can be found here:
www.LimitlessResilienceKit.com
Or scan this QR Code:

Honor Your Losses, Love, and Live Life Limitlessly,

PART II
THE BEGINNING OF BUD

17

LIFE ON THE FARM

Artie and I lived on the family ranch and farm in the middle of nowhere Wisconsin. After being married for two years, I was 8½ months pregnant. July was certainly not the ideal time to be having a baby. In fact, it was a downright inconvenient time in the farming world.

Harvest time on a huge, busy farm was endless and exhausting in and of itself. In addition, there was alfalfa to cut and bail. Compounding the situation and making it even worse was the fact that it was hot, humid, and miserable. For me, it was not an enjoyable time to be over eight months pregnant.

Labor and Delivery

Since this was long before cell phones, Artie forced me to sit in a Ford F-250 four-wheel-drive pickup truck at the end of a field he and his father were harvesting at the time I went into labor. Artie checked on me when the grain bin got full on the combine harvester. Basing his choice to work on the time between my labor pains, Artie continued the harvest.

Finally, it was time to go to the hospital, so Artie shutdown the combine and drove us into the nearby small-town hospital. After another twelve hours of labor, the baby made its appearance on July 17.

The doctor was thrilled with the delivery and made his announcement. "He's a healthy baby boy, with all the appropriate parts. He weighs 8 pounds 7¾ ounces."

This news was an unbelievable relief for me; I had done everything within my power to produce a healthy child. The risks of birth defects from excessive alcohol consumption scared the crap out of me, so I had quit drinking alcohol four months before I got pregnant. People, including my doctor, had no idea I had been a heavy drinker for twelve years. Surprisingly, I started drinking at age seven, but that is another story for The Next Day Came Trilogy, Book Three: *Kathy*. The entire time I was pregnant, I also ate healthy and was not around anyone who smoked.

In addition, I took prenatal vitamins daily, along with an extra calcium pill in hopes the baby would have strong healthy teeth. This was a concern for me as everyone in my family suffered with soft teeth and innumerable trips to the dentist office. Both my parents and two of my siblings needed false teeth in their twenties. My goal was to prevent this phenomenon in my child, if possible.

Family Names and Real Names

Artie's family believed Artie and I would never have a boy as he had five sisters. He was the only male heir to his family name. When the baby was born a boy, Artie's family insisted he carry the full family name. My little baby would be the IV, Artie was the III, but called Artie. Artie's father was Jr. and called by the full family name, which for legal protections, I will not include here. Even though Bud carried the family name, I never called him by that name. He was my little Buddy. Period. Final.

When Bud was big enough, he wore a western belt in his jeans with a buckle that said, *Buddy*! He wore this belt every day. He loved to dress as a cowboy, always neat, with his shirt tucked in so his buckle showed proudly. As he matured, he shortened his name to Bud.

Buddy Belt Buckle

Funny Name Stories

Later, as Bud became an adult, he would say to women, *This Bud's for you!* after the Budweiser beer commercials on television. I asked him if that worked for getting dates, and Bud replied, "Oh yes it does, Mother." We laughed about it together.

When Bud was mad or disappointed with his father, he decided to change his last name. Melinda said, "Bud, you should change your last name to Lite, then you would be Bud Lite."

They both laughed and laughed about that one. Thankfully, he never followed through, but I digress.

Medical Issues

Unfortunately, Bud was born with a condition called A-B-O incompatibility and became jaundiced. The nurses placed him under an ultraviolet (UV) light in the nursery. The scary part for me was they blindfolded his eyes to protect them from injury.

This was traumatic for me as a nineteen-year-old new mother. Exhausted from childbirth, in pain, and alone at the hospital, I panicked. Artie had gone back to the farm and continued the harvest. God knows that farm and those cattle came before anything in our marriage.

During this UV light treatment, Bud was nude, except for the eye patches. Suddenly, the nurse slammed into my room in a state of excitement.

"Come quickly!"

After twenty-four hours of intense labor, I was not going anywhere quickly. It turned out all the excitement was that Bud peed so forcefully that his urine hit the UV light four feet above him. The nurses found that hilarious. Afterwards, they took a medical mask and turned it sideways to make a little string bikini to prevent future urine incidents.

Bud and I stayed in the hospital, together, for nine days.

Family Support

While Bud and I were still in the hospital, Lynette, Mac (8), her three children (age 2 and 5-month twins) and their stepfather drove from Utah to the farm in Wisconsin. Looking back, I wonder where Melinda was. I appreciated that Lynette came to help me with Bud. She knew more about babies than I did, and at least I was not alone with a newborn baby in the middle of nowhere Wisconsin.

Even more helpful, Lynette was willing and able to make meals for the guys working on the farm. The stepfather helped on the farm, as he had grown up on a farm and still loved the work. The boys loved all the farm animals, especially the dogs and cats. We had three dogs and twenty-five cats. Not allowed to live in the house, per Artie, they all slept in an old chicken coop building that I fixed with straw bales for comfort and warmth at night.

Lynette and the rest of the family stayed with us for a month. I was scared and sad when everyone went home. This put tremendous responsibility on me to be alone with newborn baby Bud. With no help from Artie or his family, I did the best I could.

Artie finished the harvest, but on the farm, there was never time off. Next, it was time to work the fields. Then, the alfalfa needed cut, raked, and bailed again. After that, it would be time to plant the winter crops. Then, the two hundred head of cattle came home from the summer pastures. It was never ending—24 hours a day, 7 days a week, 365 days a year.

Doctors' Advice

By the time Bud was eight weeks old, I was at my wits' end. He slept two hours at a time, day and night. Bud woke me up screaming. I changed his diapers and clothing as needed, and then I fed him another bottle of formula. Finally, I reached the end of my sanity. I took him to the doctor in town.

"It's him or me, but one of us is going," I said.

The doctor thought this hilarious. "Bud is hungry. Feed him a jar of solid baby food or baby cereal every time you feed him a bottle of formula," the doctor said.

From that point forward, Bud ate baby cereal and four jars of baby food a day, along with eight or more bottles of formula. Thankfully, he slept for four or five hours at a time. Bud was always hungry. This fact never changed throughout his entire life.

The Toddler Years

Bud grew quickly, and I missed my work on the farm, so as soon as Bud was able to sit up by himself, I took him with me on the farm equipment and in the trucks. Bud also helped me mow the grass on the riding lawn mower. That project took all day. Bud walked before he was nine months old. He was curious and loved to learn new things.

As a toddler, I bought supplies and made Bud a fenced-in yard off the front porch. He had a safe place and was able to play outside without risk of injury by farm machinery or animals. He loved the freedom of his own place to play, but farm life was lonely for him. Our nearest neighbors with small children were miles away.

Bud and I spent all our time together. This was the start of a longtime friendship for us. We seldom saw his father as the farm kept him occupied. Artie had no interest in the baby or the toddler as he grew. Somehow, I did not see these red flags.

Bud on the Combine

When Artie and I married, I thought we would be married for life. This was the reason that no negative thoughts came to me about Artie's lack of interest in his child. My horrible relationship with Torrance, the Marine Corps Sergeant, had caused me to develop a low opinion of men as fathers. Besides, Artie was always busy with the farm, cattle, and equipment.

Vacation Time

When Bud was just over two and a half years old, he and I took a trip to Bighorn, Utah, to babysit Lynette's children and Mac. Melinda, her new husband—the stepfather, and Lynette went on a huge vacation to Hawaii for a week. I felt like I got the short end of the stick, but Bud had his first trip on an airplane, and he loved it. He always watched the sky and pointed out airplanes to me.

At the beginning of the trip, I was just over five months pregnant with our second child. However, I looked extremely pregnant. I had skinny legs and a big round stomach. I thought I looked like a bird. Looking back, it is strange how the image of a bird popped into my head, as birds have become so significant in my life.

When I got pregnant, I weighed 103 pounds, and gained 19 pounds during the pregnancy. People stopped me and inquired about the number of weeks *past due* I was.

I laughed and said, "I am only 5 months pregnant."

Shocked, they could not believe I was not about to drop the baby right on the spot.

Funny Boat Trip

One day while the two older boys went to school, I decided to take Bud and the twins on a commercial boat ride at the nearby lake. They were all excited when I told them, "We are going to see if the *damn thing* is still running," as Grandma Dove called it.

After I bought our tickets, the three cousins, all the same age with white-blonde hair, and me waddled onto a boat loaded with vacationing seniors. Suddenly, the boat grew exceptionally quiet, and the people glared at me. Eventually, one of the senior ladies became brave enough to approach me.

"Excuse me young lady, but are they triplets? Are you pregnant again?" The lady asked, pointing at the matching children.

Realizing she thought me crazy, I laughed aloud. "No, he is mine. The two girls are his cousins."

The boat passengers were relieved. Had I been a little quicker on my feet, I could have messed with their minds for the entire boat ride, but I was too tired. The kids loved the boat ride and were all well-behaved. It was sweet and enjoyable to be out in the sunshine and fresh air.

Bud experienced his first swimming pool at Melinda's house. He had absolutely no fear of the water. He learned to swim right away and spent every waking minute he could in the pool with the other children. When we arrived back at the farm, Bud missed the pool, so I looked around the farm for what I needed. I made him a makeshift swimming pool out of a cattle water tank. He swam and played in it all summer; he loved it.

Not So Funny Story

One of my favorite stories happened just before the new baby was born. At his grandparents' home, Bud was busy playing and spinning circles, like soon to be three-year old toddlers do. He fell and split his forehead open on the corner of the coffee table. Blood was everywhere.

With no help from Artie's family, even though my stomach was extremely huge at eight months pregnant, I drove the two blocks to the emergency room. Bud insisted I hold him in my arms. Yes, this was before car seat requirements.

Blood quickly filled Bud's eyes as I drove. "Mom, am I going to Kevin?" He did not say heaven.

"No, you're going to get some stitches," I laughed.

Unfortunately, he did the exact same thing the next weekend. This time, he tore the stitches out. I was right back in the same emergency room. The same doctor re-stitched where he had stitched the weekend before. If this happened today, the medical staff would have called child services on me and had me arrested.

The Beginning of the End

A huge disappointment happened when I was eight months pregnant. Artie decided to sell farm supplies to fellow farmers and ranchers to supplement our farm income, or so he said. He went to appointments around the county and made sales calls. Oddly, he started to come home later and later.

One night at midnight, I suddenly realized farmers would not be buying farm supplies at that time of night, as farmers typically go to bed and get up early. I woke Bud up and loaded him into the car. Together, we went to look for Artie, and I found his car at a bar in the neighboring town of Scully.

Imagine a very pregnant woman walking into a bar full of drunk patrons, loud music, smoke, and alcohol consumption after midnight, holding the hand of a toddler in his pajamas—only to find her husband dancing with another woman. By the look on his face, Artie, embraced in a dance with a young dark-haired girl, did not expect to see Bud and me in the bar. With one hand holding Bud's little hand and the other holding my huge pregnant stomach, I kicked Artie in the ass with my foot, as that was the easiest thing to do at the time.

I could not believe Artie trashed his marriage vows with another woman, much less while I was pregnant with his child. What an ass. I walked proudly out of the bar, put Bud into the car, and drove back to the farm. Yes, as a hormone-overloaded pregnant woman, I cried all the way home, but when I got home, Bud and I went to bed.

Later, I learned the dark-haired girl's name was Gail, per the local chatter. People gossiped about this bar episode for a long time. Our marriage truthfully ended that night. I tried to forgive and forget, but Artie had no interest in saving our marriage or his family.

Artie's cheating finished our marriage as far as I was concerned. I could see no way to fix it.

I also saw no other choice than to stay on the farm until the baby was born. I had insurance and my local doctor who had delivered Bud, so I was confident with that situation. However, the home front was not a copasetic situation; it was more like a war zone. I decided I was not leaving and giving Artie the satisfaction and convenience of having his new girlfriend at our home whenever he pleased. I would make it as inconvenient as possible.

18

SURPRISE BIRTHDAY PRESENT

The baby was born on Bud's third birthday, July 17, and I named him Jeffrey. Bud was so excited to have a little brother. He could not wait for Jeffrey to come home from the hospital. Unfortunately, that took over a week; I tell that story in my first book, *Jeffrey*. Bud was a little disappointed when the baby arrived and did not get up and play with him. However, Bud brought toys for Jeffrey and placed them in his crib, hoping he would start playing soon. It was so cute.

Surgery Required

About a month after Jeffrey was born, Bud woke up from a nap and his abdomen was swollen. I called the doctor, and he told me to bring Bud in at once. The doctor took one look at Bud and diagnosed him with a hernia. Bud needed surgery, so they scheduled it for the following day.

Once again, farming took priority for Artie. Besides, he was busy with his affair. As usual, Artie's family, who lived in the same town as the hospital, was not able to help me. So, I took both boys to the hospital for Bud's surgery. There I was with a three-year-old about to have major surgery and a newborn baby.

Jeffrey and I spent the night with Bud in the hospital. This was quite an adventure. There was little rest for me during the night as I comforted Bud as he recovered from major surgery. I slept when I could in a hospital chair. Jeffrey was in a portable crib or baby seat. Thankfully, Bud did well with the surgery and healed quickly.

The three of us went back to the farm, and I tried to put our lives back together. Artie still had no interest in saving our marriage. Thankfully, Jeffrey was a wonderful baby, unlike Bud had been. He ate, slept, and repeated; Jeffrey was a content baby. I doubt I would have made it through this horrendous time if Jeffrey had been like Bud, a starving and unhappy baby. Bud continued to scrutinize Jeffrey's growth and changes every day.

The End of a Marriage

Unfortunately, Artie and I took part in daily fights and arguments. Jeffrey must have felt that World War III went on over his head as he laid in my arms nursing. I never figured out how he was such a peaceful, loving child after the first eight months of his life. I had read that a baby picked up the atmosphere of their environment. Luckily, Jeffrey did not become an argumentative, unhappy baby.

Artie continued his affair and disappearances late into the night. This abuse was pushing me over the edge, but I needed to stay for a surgery I was supposed to have right after Jeffrey's birth. Unfortunately, because of the trauma caused during Jeffrey's delivery, the doctor would not do the surgery until December. So, I waited in a living hell, but I vowed to be out of there upon my recovery.

Life proved to be ironic for me. Artie and his Lutheran family, who attended church every Sunday come hell or high water, considered me a heathen because my parents neglected to raise me with any organized religion. Yet, it was Artie who had the affair and destroyed our family. It was Artie who was too afraid to tell his parents I was leaving with their grandsons.

When I told Artie's parents that I was leaving and why, they went crazy on me.

Artie's mother, Reba yelled, "My son would never have an affair! We raised him in the church. You will never make it on your own. You'll be back in six weeks."

I never went back. I bought an airline ticket for Lynette, and when she was on her way, Bud, Jeffrey, and I loaded into the car. We drove the hour to Red Water airport to pick her up. Together, the four of us headed to Bighorn, Utah. While we were driving, Artie's parents went as far as to call Melinda and tell her a thing or two.

Reba said, "Melinda, if you had raised Kathy correctly in the church, this would never have happened. Who does she think she is, taking our grandchildren away?"

Later, when Reba found out the truth about Artie's affair, she called Melinda back and apologized.

I had no idea how I would make it on my own, but I vowed to make a life for Bud, Jeffrey, and me, just to spite Artie and his family. This was when we became *The Three Musketeers*. Artie and Gail later married but never had children of their own. They had little to no contact with Bud and Jeffrey as they grew into adults. There were no birthday cards, Christmas presents, or child support payments.

The Three Musketeers

A new adventure in life began for *The Three Musketeers*. I was twenty-three, Bud was three, and Jeffrey was eight months old. We quickly settled into Melinda's home, and she helped me get a job at Levi Strauss & Co., a local distribution center where she worked. Lynette and I shared the babysitting of five kids, she had three under the age of five. My grandmother Dove helped when our schedules overlapped.

Bud was highly intelligent. He loved learning from an early age. He was excellent at math and loved reading. Bud read books every night at bedtime. When Bud was four years old, he could spell, print, and write in cursive everyone's full names in our immediate family.

After a year at Levi, I wanted to find something with a future, so I decided to apply to the Fallen Meadow Police Department. I passed the written, agility, and psychological tests. They accepted me into the academy, an intensive fourteen-week academy that would require my full attention. This was a terrific opportunity; it was important that I succeed and graduate.

Fathers and Surgeries

I decided it would be best to have Bud stay with his father during the time of the academy. I flew Bud to Wisconsin and started the academy. Not long after Bud got there, I received an emergency phone call from Artie that Bud had another hernia. The doctor did not think it was safe to fly Bud home before he had surgery.

I made an impossible choice. If I left the police academy, my job would end. If I stayed, I would not be there for Bud's surgery. I was annoyed. I needed a career that offered financial security and allowed me to raise my two boys by myself, so I chose to stay at the police academy.

I trusted Artie and Gail would be there for Bud, and I had confidence in the same doctor doing the surgery again on the opposite side. The surgery went well, like the first one, but the recovery was not as easy. First off, Bud was older. Secondly, his mom was not there to comfort him. Bud complained of being in pain. It broke my heart.

The doctor later admitted that a birth defect caused both hernias. Thankfully, they healed completely. Bud never had any other problems with hernias. I made it through the police academy, graduating second in my class. Bud returned home safely from his father's house in Wisconsin.

Suddenly, the bills for the doctors and hospital started showing up at my house—no surprise. These medical facilities started to threaten me with collections for not paying the bills in full. I called them and calmly explained to them they had a huge problem. They let Artie sign for Bud's surgery, when Artie did not have legal custody of Bud and had no authority to authorize the surgery.

Furthermore, Artie had no responsibility to pay for Bud's medical expenses.

I said, "You have two choices. First, you can go after Artie for your payment since he was the one who signed your paperwork. Good luck with that as he never pays his court ordered child support. Or you can accept $25 a month until I pay the bills off. And, if you mess with my credit, I will take you to court, and put you on every television news station in Wisconsin."

They accepted the $25 a month, and I paid the bills in full, eventually.

Sleep Disorder

Bud never was a good sleeper from day number one. The rule, according to Bud, was if he was up, everybody was up. Growing up, he always had a set bedtime at 7:30 p.m. The boys could read if they were quiet. I never had a problem with the boys and their bedtime. It started early, and it became a habit.

However, I quickly learned that Bud walked and talked in his sleep. When he got up, I had to listen carefully to figure out where Bud headed and what he was going to do. For one thing, he did not always use the bathroom toilet to urinate. Sometimes he used the tub or even the garbage can under the sink.

Secondly, Bud also liked to get up in the middle of the night and eat. Of course, Bud was always hungry, even in his sleep. I had him evaluated and the doctor diagnosed him with a sleep disorder.

The doctor said, "Bud goes into a very deep REM sleep and stays there."

One night, as I enjoyed my evening drink and watched television, Bud got up, walked down the hallway in his underwear, and started out the front door. "Bud, where are you going?"

"Mom, I'm going to school."

"Bud, it's midnight."

"What?" Bud asked.

"Bud, you cannot go to school now. It is dark out. Go back to bed."

"Well, I am going to be late for school. It will be your fault, Mom."

I always wondered if he walked out the door and went to school: *What would he have done in the dark?* What a sight—a young boy walking to school in the dark, in his underwear.

Bud at 8 Years Old

This photo was one of my favorite photos of Bud, even if he was holding a soda can. Look at the smirk on his face and the sparkle in his eye. Everybody loved his hair color. Both Bud and Jeffrey were handsome dudes, with the most beautiful white-blonde hair.

One activity that helped Bud with his sleep disorder was swimming. He joined the Bighorn Swim Team in elementary school and practiced Monday through Friday after school for three hours. Swimming laps burned off excess energy and tired him out.

Bud was a good swimmer and learned the techniques quickly. He loved swim meets and the competition. We had a swim meet either in Utah or California regularly. Bud won innumerable medals and trophies during his years of competitive swimming.

A Change of Plans

As Bud matured and went through puberty, he decided he wanted to go live with his father. He spent his freshman year of high school with Artie and Gail in Tabor, Wisconsin. Things went well for a while, and Bud was happy.

Bud was excited to learn about the farm. He spent time with his father and learned how to drive the farm equipment. He also learned how to care for the cattle. He experienced all the things he would have done had he grown up on the farm.

At school, Bud even joined the Future Farmers of America (FFA). He won competitions and earned his way to the National FFA event in Big Water, Utah. Artie, Gail, and Bud made the trip to Big Water for the event. What a rarity for Artie to take a vacation from the farm. Bud did well in school academically, and he made friends.

Then, something bad happened; I have no idea what it was, as neither Bud nor Artie ever said. All I know was that their situation changed drastically. Later, Bud told me his father threw him out of the house, and he ended up living on the streets in Red Water, Wisconsin. Bud never talked about the situation or how long he was homeless.

It broke my heart that his father treated him in such a manner, and it made me incredibly angry. I hold incredible anger towards Artie. He could have had the decency to call and tell me Bud needed to move back home, or that he had no idea where his minor child was or if he was safe. I never forgave him for this injustice, along with an extensive list of other things. This man had no idea what it meant to be a parent.

Bud finally called and asked, "Mom, can I come home?"

"Of course," I said. "Yes!"

Education and Changes

Bud returned home the summer before his sophomore year of high school. It was fantastic to have him home with us. I had missed him. Surprisingly to Bud, during his stay at his father's house, Jeffrey and I had moved to Turquoise, Arizona, after a job promotion and transfer with Box Trucking.

Bud attended high school in Turquoise but was extremely bored

I do not know Bud's actual IQ, as I never had it evaluated, but Bud was highly intelligent. In fact, even into high school, he was on the National Honor Roll.

"Mom, they are not teaching me anything at this school. It is so boring," he said.

Two things happened when Bud turned sixteen. First, he got his driver's license. He really wanted to get his motorcycle license, but I would not sign for that until he passed a safety class. Bud found a class in McCulloch, Arizona. Together, we took the course, and both passed it. Bud and I enjoyed spending time together and learning to ride.

Secondly, just after Bud turned sixteen, he dropped out of high school; parent's permission not needed. Bud got a job at a local truck stop. He worked hard, but he quickly realized making minimum wage was not what he wanted to do with his life.

Throughout their growing-up years, I taught Bud and Jeffrey the value of an education. I told them they could do anything and go anywhere with an education. Contrary to my experiences in childhood, I always told both Bud and Jeffrey how smart they were. My life's focus was to give them chances to succeed in life.

They saw how hard I worked to raise them as a single parent with only a high school education. This was why I made The Promise to those two little boys that when they grew up, I would go to college and get an easier job in my later years. Determined, Bud looked for a career with a future, along with a way to pay for college.

Military Requirements

"I went down to the recruiters and found out what it takes to join the military, Mom."

"Okay Bud, what did you learn?"

"The recruiter told me that I need to be seventeen and have my general education degree (GED), along with fifteen college credits. Mom, will you go over to the community college with me?"

"Absolutely! Let's go."

The registry office told Bud that he would need to speak to the Dean of the school. When we got into her office, Bud explained his situation and what he wanted to do. The Dean was skeptical.

"Why would I let you take fifteen credits? You just dropped out of high school."

I chimed in, "Ma'am, Bud is highly intelligent. High school bored him. He wants to join the military. Trust me, Bud will do what it takes. He will not be a problem."

"Okay young man, but you'd better not give me any problems."

"No ma'am, I will not."

"Thank you," I said.

Bud signed up for college. By the time he turned seventeen, he had completed eighteen college credits and passed his GED test requirements for the U.S. Navy.

19

BUD JOINED THE NAVY

Bud's plans came to fruition on his seventeenth birthday. He called the recruiter and scheduled a meeting at our home. When the recruiter showed up, Bud verified that he had met all the requirements to be able to enlist in the U.S. Navy. I never dreamed that this was how we would celebrate his birthday. As his mother, I dreamed Bud would go off to college and become an attorney or an accountant to live up to that *big name* Artie's family gave him at birth. Bud still could, but for now, he was taking his own route.

Bud wanted to join the U.S. Navy, so because he was a minor by law, I signed the contracts. Gratefully, the divorce courts in Wisconsin awarded me complete control, care, and custody of both Bud and Jeffrey back in 1983 when I divorced Artie. This uncomplicated the situation, as we did not have to contact Artie about Bud's decision to join the Navy. Nonetheless, the subject of Artie proved interesting later in Bud's Navy career, but that will be another story.

With the contracts signed, the Navy scheduled Bud and flew him down to Phoenix, Arizona to the military entrance processing station (MEPS) for testing. Bud scored high and had innumerable career choices. His first choice was to be on a submarine. Sadly, Bud, at six-foot-three-inches, was too tall for submarines. He would be constantly hitting his head, they told him. With that, Bud decided to become an aviation structural mechanic, responsible for the maintenance of aircraft fuselage, wing airfoils, fixed and moveable surfaces, and flight controls—no height requirements there.

Author's Note: Mothers should also attend this MEPS center, especially when seventeen-year-old boys make life-altering, long-term decisions. *Just sayin.'*

Bud Story

From birth, Bud was always hungry, and he never grew out of it. Just before Bud left for the Navy, Bud, his girlfriend Jodi, Renea, and I shared an enjoyable dinner. Bud finished his dinner and everyone else's food at our table. Incredibly, Bud asked a total stranger at the next table for her food.

"Excuse me, ma'am, are you going to finish that food?" Bud asked.

She looked at him as shocked as the rest of us. "No, I'm done," she said, as she handed her plate over to him.

"Bud, are you kidding me? You are going to eat a total stranger's food?" Mortified, I would have hidden under the table if I could have.

"Mom, the restaurant will throw it away. There is nothing wrong with the food."

"Oh my God, Bud, you are unbelievable. I would have bought you more food."

"Why waste perfectly decent food, Mom?"

"Bud, you don't even know who that person is."

"She looks like a respectable person, Mom. She didn't spit on it or slobber in it. I watched her."

"Bud, you are something else."

"Mom, don't worry about it. There is nothing wrong with the food."

Off to the Navy

Bud left for basic training in December as planned. Incredibly, I could not talk to him again until after his graduation in February of 1997. That was dependent on attending his graduation; of course, I would attend as his mother. This was quite a change for Bud and me. We had talked and shared things every single day since Jeffrey went off to boarding school in Plateau, Washington. I could only talk to Jeffrey on Sunday during this time.

Since I would not see Bud, I arranged to spend Christmas 1996 in Tahiti with Jeffrey. I bought tickets and made it clear to Jeffrey this was his Christmas present, period. He was not to ask for anything else. I hoped there was a celebration at basic training for Bud and the other recruits, but I was not sure what would happen for him.

I worried about Bud's weird sleeping disorder and if he would be okay in the Navy. Stress and lack of sleep intensified his condition. If Bud walked and talked in his sleep, would the Navy boot him out and send him home? Since Bud did not show up on my doorstep, I assumed it must not have been a problem. The first thing I heard from him was when I received a handwritten formal invitation to his graduation. I was so excited.

I looked back on all the challenges I faced as a single parent, raising two boys by myself. My sons and I stuck together and worked as a team. We were The *Three Musketeers*. It was a blessing indeed to have one son successfully on his way into the world.

Off to Graduation

Bud met his girlfriend, Jodi, in Turquoise, Arizona, before he left for basic training. They wrote each other throughout his time in basic.

Since they were still together, I invited her to attend his graduation with me.

I grew up in Illinois in a small town called Pluto. Jodi and I flew into Chicago and spent two days visiting with my friends and even saw my childhood home. On the day of the graduation, we drove to the base. It had been a long time since I had been on a military installation.

Sailors escorted us to a section with the best viewing and photo opportunities of Bud when his group marched in and took their position for the graduation. Jodi and I were immensely proud of Bud and his achievements. The graduation was spectacular.

After the extensive graduation ceremony, we were able to talk to Bud for exactly three hours. *Unbelievable!* After fourteen weeks and a flight to Chicago, all the U.S. Navy allowed us were those three hours to talk with Bud. *Really?* Anyway, we went to a little restaurant for food and talked.

Bud Story

The first words out of Bud's mouth when he got to talk to me were, "Mom, the Navy lied to me."

"What do you mean Bud? How did the Navy lie to you?"

"Remember, the Navy told me I could have all the food I wanted to eat?"

"Yes, Bud. Didn't you get to eat?"

"Well, yes, but they didn't say I only had twenty minutes to eat the food. I had to eat fast!"

I laughed so hard I almost peed myself. It was all about food for Bud. When I could talk again, I asked Bud the question I had waited fourteen weeks to ask. A week after Bud left, I received a box in the mail from the U.S. Navy. It had Bud's clothing, shoes, and airline tickets for seventeen other people I had never heard of.

"Bud, why did I receive a box with your stuff and seventeen strangers' airline tickets?"

"When you dropped me off at the airport, I organized

everybody and their luggage. We were ready when the plane arrived. The Navy guys put me in charge. I ended up with everybody's tickets. When we got here, we could not keep our civilian clothes, so they sent it all to you, I guess."

"They put you, a seventeen-year-old boy, in charge of grown men?" I asked.

Bud smiled from ear to ear. "Yep, they did!"

"Wow, you should be proud of yourself, Bud. I know I am," I beamed.

Jodi said, "Wow Bud, I'm proud of you too. I knew you could do it. Your mom sure missed you. We spend a lot of time together."

"Jodi, I'm glad you are getting to know each other. She's a great lady."

"Yes, she is," Jodi replied.

Aviation Mechanic School

Unfortunately, the time was up, and Bud had to go. His group was leaving in two hours on a bus for Florida and aviation mechanic school. It was so amazing to see him and talk to him. Bud had changed so much. I think Bud had grown taller. He certainly was skinnier, but the most obvious thing to me was that he had matured. We hugged, said *I love you*, and our goodbyes.

"Learn lots of stuff, Bud."

"I will. Love you, Mom!"

I gave him one more hug and kiss goodbye. "I love you too, Bud. Make us proud!"

This photo, taken in -15° Fahrenheit, was a special moment for a mother and her son.

Kathy and Bud

I gave Bud and Jodi five minutes together before Bud had to leave. Although it was a distinctly short visit with Bud, I was proud to see him graduate. I knew in my heart that Bud was going places in his life. He would certainly be much happier in the warmth of Florida.

Surprise Wedding

Bud attended school in Florida for twelve weeks and learned how to repair airplane hydraulics and mechanics. During his time in school, Bud and Jodi decided to get married. Jeffrey and I were both surprised by this decision; we were sad that Jeffrey had to miss the wedding.

Jodi and I flew to Florida and spent a couple of days exploring the area and beaches. Bud had Friday and the weekend off, so we headed to the county courthouse where Bud and Jodi were married in a service performed by a lady judge.

Half a dozen of Bud's Navy friends went to the courthouse for the wedding, and then we all went to a fancy dinner, as the boys called it. Since they were all young men learning to be sailors, I paid the bill. That really surprised them. I got the feeling they never had these kinds of special things happen in their lives. It was a wonderful experience for all of us.

Discouraging Information

Hanging out with these young men made me aware that they all smoked. This upset me that Bud had taken up smoking cigarettes. When I asked what was going on, these young men had quite a story to tell. Bud, along with his Navy friends, shared the same story.

"The whole thing is kind of discouraging, Mom. The military actually encourages newly enlisted personnel to smoke."

"What? Are you kidding me?" I asked.

A friend of Bud's spoke, "No, it's true, ma'am. Leaders announce to everybody: *if you got 'em, smoke 'em*."

"It doesn't mean you have to smoke, does it?"

"Mom, if you don't smoke, they require you to keep working. If you smoke, you get a twenty-minute break."

They all shook their heads in agreement.

"Well, that's stupid. Why would they do that?" I asked.

"I don't know, Mom, but it wasn't long before most everybody smoked. Why would you keep working while other people were smoking and laughing at you working?"

This was discouraging news to me. I was not happy at all about it.

The problem arose later when any military person decided to quit smoking. For example, there is a known anti-depressant that helps people quit smoking in less than thirty days. However, military personnel cannot get this prescription because it will go on their military record as *seeking psychiatric care*. This diagnosis would affect their chance of promotion or re-enlistment in the future. What a crazy practice indeed.

Wedding Present

I let Bud and Joni spend the weekend together for a short honeymoon, as it were. I enjoyed a peaceful quiet time on the beach by myself and read a book. What a pleasant break for me to sit on a beautiful sandy beach instead of bouncing down a highway in a big truck. Since Bud still had four weeks before he finished his aircraft training, Jodi could not stay, even though she was a military dependent after the wedding. Jodi and I returned to Arizona.

Graduation Present

Bud graduated first in his aviation structural mechanic/hydraulic class and received an early promotion. He received his orders to NAS Cecil Field serving with VFA-106 in Florida. For a graduation present, I decided to give Bud my 1969 Chevy Nova. The car was in great shape, but the color had faded in the Arizona sun, so I had it painted a beautiful dark blue.

I told Bud, "I will drive the Nova and help move Jodi, along with your poodle, Daz, to Florida."

Settled In

The trip went well to Florida. As soon as we arrived, Jodi and I found a little house for them to rent a couple of miles from the naval base. We found basic furniture for the house. They had a cute little set up, and Jodi quickly found a job.

I loved Florida, especially the beach area. I spent time at the beach and found a peace near the water I had only experienced when I lived in Hawaii. As I sat by the ocean, I decided when Jeffrey returned home from boarding school, we would move to Florida. I wished Bud and Jodi the best of luck and flew home to Arizona and back to work.

Stationed in Florida, Bud learned more about cars. By the next time I saw the Nova, Bud and his Navy friends had built a race car motor for the car. The Nova was loud and jacked up with big slick tires in the back and narrow tires in the front. It was like riding in a vibrating machine. Bud was proud of all the things he had learned in the Navy.

Bud's 1969 Nova and Engine He Built

20

NAVY SHOWCASES

I was never in the Navy, or any military for that matter, so if I get any of this wrong, I apologize. During Bud's active-duty time in the service, he learned how to repair the F-18 A/B/C/D Hornet fighter jet. The Navy stationed Bud to shore duty at NAS Cecil Field, VFA-106 in Florida for two years.

Navy Aircraft Carrier—U.S.S. Enterprise

Stationed to shore duty all four years of his active-duty commitment, Bud volunteered to go to sea on temporary duty (TDY) twice. On his first TDY, he went for a three-week assignment on the U.S.S. Enterprise, the oldest aircraft carrier used by the Navy.

Bud on the U.S.S. Enterprise Aircraft Carrier

Bud was so excited, as this was the aircraft carrier used in the filming of the movie *Top Gun*, starring Tom Cruise, one of our all-time favorite movies.

Bud Story

Surprisingly—or not—Bud had one major complaint about being on the U.S.S. Enterprise. Yes, you guessed it. The food service was slow. He complained he could not get enough to eat. Bud was always a little dramatic when it came to food.

"Mom, I would have to wait so long in line to get food that there was little time left to eat. It was like being back in basic training. I almost starved to death those three weeks. I cannot imagine being on an aircraft carrier for six months. I would starve to death."

As seen in the movie *Top Gun*, Bud was a plane captain. He performed the final safety check before the plane launched. Bud saluted, verifying to the pilot that he was good to go, and the F-18 catapulted off the deck of the U.S.S. Enterprise.

Bud Ready for Work on Deck

I learned from Bud that each job on the deck of the aircraft carrier was color coded by the shirt they wore. He wore a brown shirt in this photo. Later, in the reserves, he wore a green shirt. I had no idea what color went with which job, but I later met a woman who loaded the bombs onto the planes before they left for bombing missions. She wore red shirts, and she said, *Kaboom!*

This next photo showed Bud standing where Tom Cruise stood when, as Maverick, he threw Goose's dog tags off the ship after Goose had died in the flat spin of their aircraft. This was one of Bud's favorite photos.

DR. KD WAGNER

Bud Where Maverick Threw Goose's Dog Tags

Navy Accolades

Another exceptional story of Bud's Navy career happened when unexpectedly the Blue Angels landed at NAS Cecil Field. The Blue Angels are the U.S. Navy's prestigious elite squadron that performs at air shows and events around the world. Having been to a dozen air shows, I can attest that the pre-show was quite spectacular, in and of itself, along with the roar and speed of the fancy blue jets.

Before the jets take off in a display of military precision and showmanship, the pilots and the ground crew check out the jets' readiness for flight as a synchronized unit. After the pilots climb aboard the jets, they start the jets, with the ground crew supervision, and go through another precise process, verifying that everything is *good to go*. Then, the pilot and ground crew exchange a salute with each other, the jets parade past the crowd, turn around at the end of the runway, and roar off into the sky.

Bud called, all excited, to tell me the story. "I had an unbelievable day, Mom. We had no idea, but the Blue Angels landed at NAS Cecil Field. Suddenly, everybody was yelling and running out of the hanger. The jets landed in formation and taxied over to our area. It was so cool."

"Wow, Bud, I bet you were the most excited of everybody."

"Mom, really? Okay, I might have been. Anyway, we found out that they were in the area for a fly over at a football game later that night in Florida."

Blue Angels

"Wow Bud, what a thrill. I wish I could have been there to see them in person. You could have introduced me."

"Sure Mom, you know I would have. Anyway, it got even better. One of the pilots needed to leave early, and he pointed at me. He said, 'Sailor, come plane captain for me. I have to leave.' I was like: *Holy crap, me?*"

"Bud, why wouldn't he pick you?"

"Mom, I said, 'Sir, I don't know how to do all that fancy stuff you do for your show.' He told me,

'Don't worry, just do what you're supposed to do, and we'll be fine.' I was so nervous."

"Wow Bud, how exciting. How did it go?"

"I followed my training and launched the plane," Bud said.

"Well, I'm proud of you. I wish I could have seen you launching a plane and him taking off."

I thought I had heard the end of the story, but Bud called again the next week all excited again.

"Mom, remember my story of last week with the Blue Angels?"

"Yes, why?"

"I forgot all about it until today when I heard someone yelling my name. 'Bud, get your butt into the commander's office, now!' I thought for sure I was in big trouble."

"Why would you be in trouble?"

"Mom, you don't get told to get your butt to the commander's office for a good thing. When I got there, the commander handed me a paper. Amazing! It was a handwritten letter from that Blue Angel pilot. He sent it to my commander about me."

"What did it say?"

The pilot wrote: *Bud was very professional, and his plane captain skills were excellent. He did an outstanding job. His professionalism was better than I have seen in a long time. This young man will go far in the Navy.*

This letter was one of Bud's proudest moments in the Navy—mine too. Writing this, still today, makes me tear up. The Navy placed the letter in Bud's permanent Navy file. Bud received the Plane Captain of the Quarter award while serving at NAS Cecil Field VFA-106.

Florida Story

Bud learned how to surf in Hawaii as a child and continued to surf while stationed in Florida. When Jeffrey and I moved to Florida in 1997, we lived at the beach, two blocks from the Atlantic Ocean. If the waves were good, the boys wanted to be surfing. Bud, his Navy friends Mark and Farris, and Jeffrey spent time together. They worked on cars, surfed, danced, drank beer, and played cards.

In fact, Bud and Jeffrey were on the news one year when they were crazy enough to go surfing while a hurricane passed off the eastern shore of Florida. Their surfing was harder on me watching from the beach than on them fighting the waves and the currents. They were having a blast while I panicked. I was not happy, but the news people loved it. Truthfully, we were all crazy to be out in the middle of a hurricane.

In January 1999, Jeffrey and I left Florida and moved over two thousand miles back to somewhere I had no desire to be to take care of Melinda in Bighorn, Utah. She needed open-heart surgery to replace a defective heart valve. Two of her children lived in the same town with Melinda, but neither would support her during her recovery, so Melinda called me.

Changes for Bud

In November 1999, Bud and Jodi divorced. They were married for two years, but I guess it did not work out for them. I never heard the story of why they broke up, nor did I ask. Jodi and I lost track of each other, but when Facebook came out, Jodi found me, and we continued to stay in touch with each other. We have remained friends and see each other occasionally, as we both live in Florida.

The Navy decided to move Bud in September of 1999. There were nationwide base closings, and one of them was NAS Cecil Field. Years later, the base became a private airport. Bud was one of the last Navy people to leave the base.

Bud's next assignment was NAS Patuxent River in Maryland. He stayed there for the remaining two years of his active-duty commitment. During his time there, the Navy selected and sent him to Boeing for F/A–18 E/F Organizational Level maintenance school. This was quite an honor.

When Bud arrived in Maryland, he met one of his best friends, Noah. They worked in the same unit together. Through another co-worker, Bud and Noah became friends with Marisol. Later, when

Marisol left her husband, she moved in with Bud and Noah. It was at this point that Bud and Marisol's longtime friendship and relationship developed while the three of them shared a home together.

Moms Come to Visit

Early in the spring of 2000, Kim and I decided to visit Bud. He insisted that we stayed with him, Marisol, and Noah. This proved to be quite an adventure for us and them. Bud had shared stories of Kim's cleaning habits with the group. Bud had never been Mr. Clean—that was his brother, Jeffrey.

Bud, Marisol, and Noah all came to the airport to pick us up. On the trip home from the airport, they told us all about the cleaning job they had worked on all week. They were so proud of their cleaning, so we never said anything, but Kim took one look at the shower and cleaned the entire bathroom and guest room again.

Boys and their toys. The garage was full of car parts. Getting into the house was like maneuvering through an obstacle course. Kim nor I had ever seen so many car parts in one place. Bud and Noah made a clean path through the garage for Kim and me; they did not want us to trip and fall.

The boys were certainly happy with their mess in the garage. Kim and I learned all kind of things about motors and cars. Bud was helping Noah restore a Mustang. Together, they were building a race car motor for Noah's car using the skills Bud had learned in the Navy.

Trip to the City

Bud, Marisol, Noah, Kim, and I all loaded up in Bud's Nova and drove to the Metro Station. They told us that parking and traffic were horrible in the city, so we took a train. It was interesting; as we got off the train, we walked up in the middle of the plaza between the White House and the Capital building.

We took a bus ride over to Arlington National Cemetery, which was serene as we watched the changing of the guard ceremony. It was incredibly sad but very precise. After that, we got on another bus called the *Hop On–Hop Off* bus. We rode around on the top of the bus and admired the Capital, White House, and dozens of monuments. We hopped off the bus and took in a couple of the museums. It was a wonderful day.

We particularly enjoyed teasing Noah. He had never met folks like Kim and me. He either enjoyed or was astonished by our stories and jokes. We could tell easily, because if he was embarrassed, his cheeks turned beet red. He was adorable.

We had a wonderful time, ate too much food, and watched horrible movies. Kim and I headed back to Utah and work. It was so fun to spend time with Bud. It really did not matter where we went, or what we did, Bud and I enjoyed talking and hanging out together.

21

BUD'S 21ST / JEFFREY'S 18TH

Bud returned to Fallen Meadow, Utah, in July of 2000 for his twenty-first birthday. In the state of Utah, a person under twenty-one years of age has a profile photo on their driver's license. Bud wanted a new license, with his photo facing forward, showing he was twenty-one years old. He planned his whole leave from the Navy around getting this new driver's license.

Both Bud and Jeffrey were born on July 17, three years apart. Since it was Bud's twenty-first birthday and Jeffrey's eighteenth birthday, I decided we would have a photo taken while Bud was home for his visit.

"Jeffrey, please do not cut your hair before Bud gets here. I want to get a family photo taken."

Of course, Jeffrey did the exact opposite and shaved his head.

Jeffrey, Kathy, and Bud

Unfortunately, the photo above was the last photo ever taken of *The Three Musketeers* together. This was the very last time we were ever together in one place.

Truth About Divorce

During this trip, Bud, Jeffrey, and I had one of the most personal conversations we ever shared. I needed to explain to them both that I found a company that went after deadbeat dads for unpaid child support. Bud and Jeffrey had no idea that their father owed me over $40,000 in unpaid child support as I had never bad-mouthed their father to them. They were shocked.

The company, Supportkids.com, would go after their father for the money he owed me. I explained to them this company worked on a percentage. If they got the money from Artie, I would get two-thirds of the amount. The company kept the rest of the money for their finder's fee and getting Artie to pay.

"Two-thirds of something is better than 100 percent of nothing," I said.

I explained to them that if their father refused to pay, he would go to jail. I suggested they think about it and let me know if they wanted me to pursue the matter.

"Mom, you have worked hard all our lives and took care of us. We both think that you deserve the money he owes you," Bud said.

"If he refuses to pay the money, then he deserves to go to jail," Jeffrey said.

"Be sure, because it could affect your relationship with your father forever," I said.

"What relationship?" They asked in unison.

"Okay, if I get the money, I'll split it with you both. One-third for each of us."

"Mom, you don't have to do that. The money is yours. You are the one who paid all the bills and provided for us. You keep it."

"Dudes, we will split it. You can spend it on whatever you want. I will let you know when I make my final decision. Thanks for understanding and letting me know how you feel."

We left it at that. It took me a while to fill out the paperwork and send it in. I did not hear from the company, and life kept me busy. In fact, I completely forgot that I had sent the paperwork in at all.

Marisol Moves Home to California

In the fall of 2000, we found out that Marisol wanted to move home to California. Bud volunteered to drive her moving truck for her. Marisol missed her family, so with sad hearts, Bud and Noah loaded her stuff up in a rental moving truck, and Bud and Marisol headed west to California.

Marisol and Bud

Later, when Kim and I saw Marisol in California, we asked her about moving home.

"Kathy, I didn't think Bud was serious about me."

"Marisol, I know Bud cares about you deeply, but he also has gone through a lot of changes in the last few years of his life. He really is still growing up. And technically, you are not free."

"That's true. I should get my divorce. I really hadn't looked at it that way. Thanks."

Kim and I privately decided Marisol had gotten tired of climbing over car parts in their garage.

Navy Aircraft Carrier—U.S.S. Harry S. Truman

When Bud returned to NAS Patuxent River, he volunteered to go TDY again. This time, he spent three weeks on the U.S.S. Harry S. Truman, the newest Navy aircraft carrier.

"Life was much better on the Truman, Mom."

"How's that Bud?"

"They served food twenty-four-seven, Mom, no lines. It was like food heaven."

"Really Bud, is it always about food for you?"

"Of course, it's all about food, Mom."

"Bud, you have never changed. You have always been hungry!"

"Okay Mom, everything was clean and fresh on the brand-new aircraft carrier too."

Active-Duty Done

Bud finished his four years of active-duty while stationed at NAS Patuxent River, Maryland, in December 2000.

One of Bud's claims to fame in the Navy was a surprise to most everyone who served in the military. The Navy stationed Bud to four years of shore duty. *That don't happen often, baby!*

Bud decided he would stay on the East Coast and get a job; however, that did not last long at all. By January of 2001, Bud decided he had enough snow and wanted to move to California. "Mom, can I come stay with you and Kim for a while?" Bud asked.

"Of course," I said. "Yes!"

Bud loaded up in the middle of a blizzard. He still had the Nova and could not drive two vehicles, so he shipped the Chevy out to California on a truck. Bud arrived safely.

It quickly became obvious that Kim and I were not the main attraction for Bud. Just after Bud's arrival, we got to meet Marisol's family. They welcomed us into their home. Kim and I did not know anyone in California except our neighbor Betty, so it was kind of them to include us in their family gatherings.

The Nova arrived safely, covered in salt from the roads. Bud had it polished up in no time. When the Chevy started up, the whole townhouse shook. Kim quickly made a rule: Bud had to idle out of the garage, then out of the townhouse complex and onto the highway. At that point, and only at that point, was he allowed to floor the engine and take off.

Navy Reserves

Bud mustered into his Navy Reserve unit at NAS Point Mugu in Oxnard, California. His trip was over an hour drive from our townhouse, depending on traffic through Poison Oak, California. Luckily, he only needed to go to the base one weekend a month as a reservist in the U. S. Navy.

Next stop, Bud signed up for the spring semester at Surf City College, where he kept a 4.0 GPA. Bud loved learning new things. This boy was going places and had lots of things he intended to do. He found a part-time job at a Sherman Williams paint store nearby and settled into his new life in California. Bud certainly had a little adjusting to do getting to know Kim and living with two women.

A Boy and His Poodle

Jake, our standard poodle, fell in love with Bud from the moment he arrived. "Jake is a chic magnet, Mom!" he said.

"Oh really? I didn't know you were looking."

"Mom! The girls love to pet him and ask me questions when I walk him on the beach."

Jake was thrilled to have a young man take him for walks and play fetch with him out in the grass. Bud could throw the ball much further than Kim or me. Bud taught Jake how to play hide-and-seek. One of us would hold Jake in the kitchen area and count to ten so Jake could not see which way people went to hide. Once everyone had hidden, Jake would find them.

Bud also taught Jake how to play tag. They ran up and down the stairs of the three-story townhouse. Bud ran halfway up the open stairs, and Jake tried to beat him to the top. Halfway up, Bud turned and jumped back down. Jake, however, could not turn as sharply as Bud. He had to go up to the landing and do a turnaround, then chase Bud the other way.

Kim and I found great fun watching Bud and Jake play. The energy and love they shared was amazing. Watching them together reminded me of Bud and Jeffrey playing when they were younger.

Bud was also the only one of us who could lift Jake up; he weighed over ninety pounds. Bud carried Jake around the house as if he was a little puppy.

Differences Between Brothers

Previously, Jeffrey had lived with Kim and me in Utah. Jeffrey was Mr. Clean. His room was immaculate. He showered twice a day, did his own laundry, and ironed his clothes to perfection.

Bud, on the other hand, was not Mr. Clean at all. Yes, he had learned and could do all the things Jeffrey did, but he chose not to do them. Kim was *Miss Clean*: she waited for Bud to get up and go downstairs for breakfast. Then, she made his bed and cleaned his room. This was a wonderful thing, except she put things away. When Bud could not find his stuff, he called me while I was driving my tractor-trailer to Texas and back instead of asking Kim.

"Mom, call Kim and ask her where she put my hat."

"No, Bud. You ask Kim where she put your hat, and she will tell you."

"Mom, you ask her; I don't know her that well yet."

"Then get to know her. Either that or put your stuff away. I am not playing phone questions for you two."

They eventually worked it out. However, Bud did call one other time. "What is a hot dish, Mom? Kim says we are having a hot dish for dinner."

"Bud, it sounds like a casserole, but I'm not sure," I said.

"Will I like it?" Bud asked.

"I have no idea, Bud. But you like food, and it sounds like food to me. Ask Kim what it is."

22

BUD'S RECOLLECTIONS - THAT DREADFUL CALL

At approximately 6:00 p.m. on March 28, 2001, I received a phone call that nobody should ever receive, much less a twenty-one-year-old young man. Thankfully, Kim was at home when the call came through. I cannot imagine what would have happened had I been alone when I received this phone call from my estranged father, Artie.

"Why are you calling me?" I asked.

When the call ended, I hung up the phone, held my head in my hands, and cried. My whole body shook.

"Bud, what is going on? Who was on the phone?" Kim asked.

"My father. He said that somebody murdered Jeffrey!" I yelled. "This will kill my mother."

"Yes Bud, but somehow, we have to tell her. There is no other way."

"Oh my God, Jeffrey?" I cried.

Author's Note: Our lives changed forever that day—Bud and Marisol's relationship and Kim and my relationship. Life itself was never the same. It took everything we had to get through the devastation of losing a son, a brother, a stepson. From this day forward, there remained times and places forever lost, only living in my memories or recollections.

23

BROTHERS LOST IN DIFFERENT WAYS

Life became complicated with the death of Jeffrey. Bud and I struggled daily with his loss. Bud's birthday came on July 17, 2001, and he wanted absolutely nothing to do with the day he had shared with his brother for eighteen years. Bud would not do it alone, at least not this year. I knew that Bud suffered from survivors' guilt. He believed he should have saved Jeffrey.

Kim and I thought we needed to help Bud move forward. We decided to give Bud a birthday week. Kim organized a surprise party—not a big affair, just a dozen friends. The party was painful for both Bud and me. Inside I was dying, but I had to be strong for Bud. People forgot that I had lost my baby, my Jeffrey, and it was his birthday too.

Surprisingly, after Jeffrey's funeral, Bud and his father developed a relationship. Artie had sold the farm and taken up truck driving. This was an odd phenomenon to me as three of my exes started truck driving after exposure to me and my truck driving career. I digress.

Bud and Artie spent time together when Artie came to California for work. They also spoke often on the phone. Interestingly, they were both men, which changed the relationship.

They were not father and son any longer. This relationship seemed to be working for the time being. Hopefully, Bud would not get his heart broken again.

Artie's Ass

The relationship between Bud and Artie was short-lived. When SupportKids.com went after Artie for the money he owed in back child support, Artie blamed Bud. God forbid Artie take any responsibility for bringing these two boys into the world. Had Artie lost his freaking mind? Bud was three years old when I divorced his father, and we left his home.

First, Bud lost his brother. Then Bud lost his father again, and this time, it was forever. That was the end of Bud's relationship with Artie. He never saw or spoke with his father again. Artie did not blame me; the only explanation I had is that Artie did not have the balls to call me. There was no doubt that tactic would have been a losing battle for Artie.

I received a check for the two-thirds of the money in back child support owed to me in July of 2001. I realized with acute anger how important this money would have been to me and my sons, had I received the money when it was due as child support. This money would have allowed me to spend more quality time with my children as they grew up, time I can never get back. Instead, I worked long hours in hard jobs to support Bud and Jeffrey as a single parent. This knowledge increased the anger and hatred I already carried in my soul.

Amazingly, the same court that refused to help me get my child support for all those years came out of the woodwork and made sure they covered their butts. Scully County family courts made me sign court documents clearing Artie of any debt owed to me for child support in arrears.

Bud said, "Mom, you can keep all the money. You worked hard to raise Jeffrey and me. I don't need any of the money."

"Bud, a promise is a promise, you know that I keep my promises."

Memorial of Jeffrey

Bud used part of his money for a tattoo to honor his brother Jeffrey. This was a heartfelt memory for Bud, so he would always have his brother with him in his heart and in spirit.

Bud also used the money to buy his 2001 Harley-Davidson. Bud loved to ride that motorcycle. If Bud was not at school, working, or the Navy, he was on that Harley.

Bud on his 2001 Harley-Davidson Motorcycle

Kim and I realized that the Honda Shadow motorcycle I rode back and forth to work from the airport still sat at my work in Fallen Meadow, Utah. We decided to bring it home since it had been sitting there since my injury in July of 2001. Since I would not go back to work there again, there was no reason to keep it in Utah.

Bud volunteered to drive the motorcycle back to California. We flew him to Fallen Meadow, and he took a cab out to my old work. Once there, he checked out the motorcycle; surprisingly, it started right up. Bud went to a local gas station, fueled up, checked the tires pressure, and headed to our townhouse in Captain Island, California.

Bud on Kim's Honda Shadow

My injury prevented me from using the motorcycle. Kim decided to take lessons and earn her motorcycle license. Once Kim got her license, she and Bud became good friends and went on rides every chance they had. Bud looked up to Kim as a mentor and hoped to emulate her in the choices she made in her education and career.

Bud also rode motorcycles with our family friend, Chad, and his Navy friend, Don. One of his favorite things to do when he was not working or going to school was to ride with the group of Harley owners who met at the Harley-Davidson store in Tropico, California. This group gathered on Saturday mornings and rode to either the mountains or the beach. This trip always included lunch, which was another attraction for Bud.

Scuba Divers

Bud and Kim also learned how to scuba dive together. When I lived in Hawaii from 1985 to 1988, I became a certified scuba diver and took dozens of dive trips. My workman's comp injury from truck driving prevented me from physical activities of any kind, so Bud and Kim shared this experience together.

Bud and Kim had to lug their equipment up and down the rocks along the beach in California to the Pacific Ocean. They even had to wear huge wet suits, as the Pacific Ocean was cold. Bud was always a gentleman and helped Kim carry her equipment. He loved scuba diving and had no fear of the water. I watched them from the beach and took photos.

Unsettled Peace

Things slowly settled down at our house. At least we found a bit of consistency in our daily lives. Bud and I continued attending college. He worked full-time for the Navy as they planned for war. I worked on The Promise to earn that degree that would afford me that easier job someday. I tried my best to be the mother Bud needed.

Unfortunately, I struggled silently with the loss of Jeffrey. His loss left a huge hole in me, my life, and family. The *Three Musketeers* were no more. *Two of a Kind* struggled every day to understand the senselessness of murder. We missed Jeffrey every single day.

Thankfully, our family began to evolve, as both Bud and I started to trust Kim. The fact that Kim stayed through a hell that most people would have run from surprised us both. Bud, Jeffrey, and I had watched dozens of people who promised they loved us and would always be there for us come and go from our life. Kim appeared to be the one who was going to stick around. She played a vital role in our lives and our survival.

I never again trusted the feeling that things were going to be okay in my life. There was always that *waiting for the other shoe to drop* feeling. Those words, never spoken, remained in my heart. There was nothing to say that would ever fix things in my life.

Besides, I learned from Jeffrey not to say things aloud because I did not want to put negative thoughts into the Universe and curse myself. Things changed whether I wanted them to or not. I constantly ran *up the down escalator* of life, and it was stuck on high speed.

One step forward—three steps back.

ACTION STEP
BACK TO THE NIGHTMARE

As I learned more about coping with my own loss, many recommended journaling and coloring as powerful tools. Portions of this book may trigger intense feelings, good or bad, while reading these stories. If this happens, write those feeling down. To help you with this, I created **The Next Day Came Trilogy Thoughts and Emotions Activity Book**, which provides space to write and color when you need to take the time to process your thoughts. Along with this resource, I am also providing you with the Limitless Resilience Checklist and How to Build Emotional Resilience Video to help you become more resilient in the face of extreme adversity.

These resources can be found here:
www.LimitlessResilienceKit.com
Or scan this QR Code:

Honor Your Losses, Love, and Live Life Limitlessly,

KDWagner

PART III
BUD'S NEVER COMING HOME

24

NAVY BLUES AND DÉJÀ VU

On August 28, 2003, at 10:10 a.m., everything in my life changed when I learned Bud was dead. Suddenly, my body felt like a tidal wave crashed into me. The wave threw me down, and I swirled underwater; I could not breathe as there was no air. As I spun around and around in the whitewater, I had no idea which way was up. How could Bud be gone? He was all I had left. The last twenty-five years of my life, I had dedicated to raising my sons. Everything I lived and worked for was gone in an instant. Déjà Vu.

How can They take my only remaining child? There cannot be that kind of evil in the world.

Breaking News

How could it even be possible that once again I was watching breaking news about my child's death on the television news, a homicide? *Secret Witness* or *Crime Stoppers* was requesting information about the felony hit-and-run on U.S. 501 Freeway. They requested information about a twenty-four-year-old Navy sailor killed in a fatal accident involving a motorcycle.

The news story revealed that Bud, a petty officer in the U.S. Navy serving on war status for his country, had returned earlier in August from his first tour in Operation Iraqi Freedom. He was on his way to the Navy base in preparation to leave for his second tour when someone fatally struck his motorcycle with their car and left the scene—a felony hit-and-run—causing death per California law.

The news announcer felt Bud deserved recognition and honor for his service and this senseless loss on American soil after being in a war zone serving our country. The reporter announced that a car ran Bud down and left him to die on a California freeway. That made the situation even more horrific and unforgivable to the reporter.

The reporter continued the story saying that one of the witnesses, a Navy commander, stayed with the victim, Bud. The second witness, an Air Force officer, chased down the fleeing vehicle and retrieved the license plate number.

What an incongruous event.

The story of Bud's death was on every local television station and in every local newspaper. The coincidences and parallels between the deaths of Bud and Jeffrey were disconcerting to me. Mercifully, one blatant difference struck me like a bolt of lightning during a hurricane. It would not be me who made the dreaded call to Melinda. Oh, hell no! I would never risk her response to the announcement of the death of a child again.

After the incredibly horrendous response that Melinda gave me when I called her at 2 a.m. to tell her that somebody murdered Jeffrey, I would not tell her that my other child, Bud, died in a homicide too. Melinda could hear the news about Bud from anybody but me. Besides, Melinda lived in Utah, so she would not get the California news.

I had not spoken to Melinda in over two years. If I had my way, I would never speak to her again. I flashed back to the night I called to tell Melinda about Jeffrey's murder. Her response, *"Are you coming to Fallen Meadow, because I need money?"* was completely atrocious. Melinda shared not one ounce of sympathy or empathy for the murder of my child, her grandchild.

Another Day of Hell

Nobody in our house slept through the night, certainly not me. My mind never stopped. *How can this be true?* We had just spent Sunday with Bud. He and his friend Don came up to Frostbite to help us move the fence. We shared a wonderful dinner with friends and family. At least we had these precious memories of Bud fresh in our minds.

People got up, showered, and got ready for the day. I did not get up. What was the point? The phone rang, and Kim answered. She talked and hung up. "Kathy, if you're not going to sleep, you might just as well get up."

"Why?" I asked.

"The Navy will be here by noon. They just called and are on their way. They are going to want to talk to you."

"Kim, what could they possibly say to me that would bring Bud back? Not a freaking thing. I do not want to talk to them or anybody."

"Get up for breakfast. Everybody else already ate and took their showers. It's your turn."

I knew it would do no good to argue with Kim. She had back up with a doctor and a nutritionist both telling her I had to eat, so I would eat whether I liked it or not.

"River will be here in a bit," Kim said.

"Your brother still wants to come up here with all this going on?"

"Yes, and there could be other people coming today, too."

Memories

It was a sweet reminder to have Lynette's grandchildren in the house. These children brought back memories of the time when my sons were young and innocent. My nephew and nieces, along with Bud and Jeffrey, grew up together. When the family got together for one occasion or the other, the kids would put on plays for the adults. It was adorable.

One of my favorite memories happened at Lynette's house. One of the twins—Lexie, I believe—was supposed to *pretend* to slap Jeffrey across the face. She forgot the pretend part and slapped Jeffrey hard. He had a handprint on his cheek, but Jeffrey never missed a line in the play; he continued with the show.

"That's show business!" Jeffrey said later.

There were so many memories of these kids growing up together. They were good, bad, and indifferent times, but they were family. Bud and Jeffrey managed to survive, grow up, and then people killed them both. Life was so unfair.

The Navy Arrived

The doorbell rang. There were two people in Navy uniforms standing outside the door. I took one look and broke down crying. Bud always looked so handsome in his uniform; I loved seeing him.

Bud at Veterans Day Event

BUD

The people at the door introduced themselves by rank and name. I only remember their first names. One was Travis, and the other was Rayna. Enlisted, the same as Bud, Rayna said that Bud and she were good friends. Travis was an officer; his position was the CACO officer. They were both coworkers of Bud's.

According to the Navy, the CACO officer provides information, resources, and assistance to the next of kin of the deceased Navy person in the event of a casualty. They explained the benefits and entitlements allowed to the next of kin. This service was helpful because the family usually did not understand the terminology, acronyms, or policies and procedures of the military.

The Meeting Began

We all sat at our dining room table. Travis explained that he had just gone through CACO training the month before and was a nervous wreck. He had his books and stuff spread out all over the table. Six adults and four children waited to hear what he had to say.

Travis spoke hesitantly, "I truly hoped and prayed I would never ever have to use this training."

I was not in a good mood and wanted nothing to do with the entire process. "That didn't work out too well for you, did it?"

"No Kathy, I can't believe I'm sitting here right now."

"Yeah, I can't believe it either."

"I am terribly sorry for your loss. Bud was a great guy. Everybody loved him."

"That's true Kathy, everybody in our unit has the utmost respect for Bud," Rayna said.

"Bud was a very special person," Travis said.

"Yes, he was," Kim said.

"Bud and I were still struggling with the loss of his brother, Jeffrey."

"Yes, the loss of his brother really affected Bud. He was heartbroken," Rayna said.

"We all are heartbroken," I said.

"Kathy, no mother should have to endure this much," Travis said.

I relaxed a little bit with these words. "Truer words were never said."

"Jeffrey's death made Bud even more determined to be the best and succeed. He certainly showed it in the Navy," Travis said.

"Bud and I made a pact to live our lives to make Jeffrey proud every day."

"I could see that. Bud was always trying to improve himself and others."

"He came in and turned our whole unit around the first day he walked in the door at NAS Point Mugu," Rayna said.

Rayna's statements made me even more proud to be Bud's mother. These two people knew and respected Bud. "How did Bud change the whole unit?" I asked.

"Before Bud came, we sat around bored; nobody did anything," Rayna said.

"That sounds pretty boring."

"It was. Bud was not going to sit around the whole time."

"That sounds like Bud. He never was one to just sit around and do nothing."

"When Bud got there, he walked right up to the commander, introduced himself, and asked, 'What do I need to do to get my next qualification or my next promotion?' It was amazing."

"Yep, that sounds exactly like Bud."

"Everybody was watching Bud. Next thing you knew, they wanted to do what Bud was doing. Now, the whole unit is fully qualified, and that is thanks to Bud," Travis said.

"Bud is something else. He takes after his Mama," I bragged.

Everyone laughed. It broke the stress for a minute.

"We are all industrious workers in our family. My father was a Marine sergeant in the Korean War. He raised us as Marines."

"It was ingrained into us all the time," Lynette said.

Kim laughed aloud at this story. "They do live by strict rules."

"I guess I raised Bud and Jeffrey with the same values."

Both Rayna and Travis were genuinely kind people. They tried their best to make us comfortable, and they made me feel a little less alone.

"Bud was truly liked; everyone admired him," Rayna said.

"He will definitely be missed. He was an extraordinarily strong young man," Travis said.

At least a stranger had not sat at my table and told me about my lost son. These two Navy people knew who Bud was. They knew what he stood for, and they respected him. It made a horrible day just a tiny bit better.

Travis went through the pages of his huge notebook one by one. I can honestly tell you I do not remember a damn thing he said. It utterly amazed me what the brain could do to protect a person from the most horrible things in the world.

It was déjà vu all over again. The same thing happened when I saw Jeffrey in the funeral home two years before. My brain would not listen to anything more about the death of my child. I was more than done.

25

FUNERAL PLANS—AGAIN

Thankfully, Kim was able to keep track of what these people said. Travis shared the information about Bud's funeral services. We learned that when military people go to war, they must make out a will and plan their funeral.

"Bud stated that he did not want to be buried at Arlington National Cemetery because people would not come to visit him," Travis said.

This brought me out of the fog for a moment. "Really? Bud said that?" I asked.

"Yes, Bud wants a burial in Riverside National Cemetery in Riverside, California, with full military honors, including a twenty-one-gun salute and echoing bugles. The flag will be folded, and presented to you, Kathy."

"Travis, I am not able to see another dead child. I just cannot do it," I said. "You can have whatever services Bud wants at the funeral home, but I will not be there."

"I completely understand. I don't know how you are breathing right now."

"Travis, you really have no idea. I refuse to see Bud in a casket,"

I said. "I saw Jeffrey wrapped in a sheet, dead. I still see Jeffrey every time I close my eyes. It is burned onto the insides of my eyelids."

This obviously shook Travis to the bone. "Oh my God, I had no idea. I am so sorry, Kathy."

"I will not do this with Bud. We just spent the weekend with Bud. We had dinner with him and Don. That is how I will remember him. Alive."

"That's fine. The Navy will take care of everything for you, I promise."

"I am so glad you had that time with Bud," Rayna said.

"That is the only way I am ever going to remember Bud."

"I can see why Bud had such respect and love for you," Rayna said.

"How would you know that?" I asked.

"We used to sit and talk about life. Bud absolutely loved you."

I was not sure who this Rayna girl was, but she truly seemed to care about Bud. I wondered to myself if they dated, but I did not ask. The girls always chased Bud. I later found out that she had a young child. Bud helped her out and played with the boy. Bud loved children, so I was not surprised by this news.

"My boys and I grew up together. We were The Three Musketeers!"

"Bud had such fond memories of growing up with you and Jeffrey. I know it is not comforting for you, but my belief tells me he is with Jeffrey. They are together," Rayna said.

"Well, I would rather they were here with me."

Rayna's eyes filled with tears. "I know. I am so sorry."

"Thanks."

About to cry himself, Travis stood up. "I am going to find out when the services will be for Bud."

He excused himself and went out onto the deck to make a phone call.

Kim decided I had better eat something. She went into the kitchen to make me a snack. This eating schedule annoyed me, but Bud deserved to be honored and respected. I could figure out how to die later.

BUD

Travis was not gone long at all. "I spoke with Riverside National Cemetery."

"That was quick," Kim said.

"Bud's service is scheduled for Wednesday, September 3, 2003, at 1:45 p.m."

My mind would not play anymore. Unfortunately, the conversation continued. Thankfully, Kim completely took over at this point as I tuned out.

Funeral Homework

"So, what do we need to do in the meantime?" Kim asked Travis.

"Can you pick out a photo of Bud that you like?"

"Sure, what else?"

"The funeral home will need some personal information about Bud for the in-memory card."

"What kind of card?" Kim asked.

"The cards are handed out at the funeral service. The card will have Bud's photo on the front. On the inside, it will open to a poem or Bible verse, whichever you prefer."

"Bud is not the Bible verse kind of guy. I'll ask Kathy later, but a poem will be better, I'm sure."

"It will also have Bud's name, date, place of birth, along with date and place of death. There will be the date, time, and location for the graveside service. It will also list the Naval Officer who will officiate the ceremony at the cemetery."

"We will gather that information for you. When do you need it?" Kim asked.

"I will come back on Monday. Can you have it ready then?"

"Okay, we will work on it this weekend. What else do you need?" Kim asked.

Travis went on to the next page of the book. "Do you want to place an obituary in the paper?"

"What do you think, Kathy?"

This just would not end. "Sure," I said.

"Bud has a lot of friends who need to know the information," Kim said.

"They need that by Saturday morning, if you want it in the Sunday paper," Travis said.

"Okay. We have help, and unfortunately, we have experience," Kim said.

"I think that Kathy has reached her limit. I will come back on Monday."

"Yes, I think we have enough homework for now."

I spoke up unexpectedly. It was hard to figure out what went through my brain, or what would come out of my mouth. "Lynette and the twins will work on the obituary and stuff you need. They wrote Jeffrey's obituary."

"Okay, you can email the obituary to me when you get it written," Travis said.

"Thank you for all your help. I know Bud would appreciate it. Can you leave us your contact information in case we have any questions?" Kim asked.

Travis and Rayna gave Kim their information. Rayna said, "If it is okay with Kathy, I would like to be a pallbearer for Bud's service. He was a good friend to me, and I wish to honor him."

"Sure. How many do we need?" Kim asked.

"Six," Travis said.

"Kathy, we can ask Wyatt. He is retired military. Bud's friend Noah is coming. He was in the Navy with Bud. Chad will be here too. Will you ask Don if he is up to it? I don't know who else," Kim said.

"Don't worry about it today. We will fill in whatever you need," Travis said.

"Okay, we will see you on Monday."

"Correct, I'll call you when we leave the base. It will be around noon again by the time we get here. It is a long drive, especially with traffic."

Kim and I walked them down to their vehicle. The Universe had dumped another big fat load of crap right into my lap. My heart broke into pieces.

BUD

Can I survive this? Will I survive this? I had no answer to those questions.

Bud's Obituary

Kim's brother River drove up the driveway as the Navy left. The twins and River, a teacher from Arizona, were an immense help with the obituary and the in-memory card. We needed to get the obituary completed ASAP.

The twins went through my computer photos and found a picture of Bud that everybody agreed upon. They cut Bud out of the photo, and this was the obituary they put together. Bud's obituary appeared in four or five newspapers, including the *Poison Oak Times*, the *Tangerine County Register*, and the *Bighorn Newspaper*.

Bud, 24, died Thursday, August 28, 2003, in California. As a Naval Reservist, assigned to VR-55 Minutemen NAS Point Mugu, CA, the Navy called Bud up after September 11, 2001, for active-duty. He took part in one overseas tour, serving in Operation Iraqi Freedom.

Bud was born in Tabor, Wisconsin on July 17, 19**. He

joined the Navy when he was 17 years old and attended basic training at the Great Lakes Naval Training Center in Illinois, and then Aviation Structural Mechanic Class A school in Florida. Afterwards, stationed at NAS Cecil Field, Florida, and NAS Patuxent River, Maryland, Bud earned an honorably discharge in December 2000 from active-duty, and joined the Navy Reserve in California.

As a full-time college student, Bud was on the Dean's Honor List and earned but did not receive an Associate of Arts Degree. In addition, Bud was an avid Harley Davidson motorcycle rider, scuba diver, surfer, golfer, and angler.

Preceded in death by his brother Jeffrey. Survived by his mother, Kathy, her partner, Kim, of Frostbite, and his father, Artie, of Wisconsin.

A service with full military honors at the Riverside National Cemetery, California, on Wednesday September 3, 2003, at 1:45 p.m.

In lieu of flowers, please send donations to: The Love Ride Foundation, Harley-Davidson of Glendale, CA.

Newspaper Article

Sharon, a reporter from the *Bighorn Newspaper* in Utah, saw the obituary and called me. This was ironic as I had delivered the *Bighorn Newspaper* all those years ago when I was first divorced to earn gas money.

She asked, "Can I interview you for an article? People need to hear the story of your sons."

Although I had no desire to speak to anyone, I did the interview for Bud. The article turned out to be genuine and heartfelt. She talked about Bud and Jeffrey growing up in Bighorn and being on the swim team for years, along with Jeffrey's loss. She even mentioned that I graduated from Bighorn High School, even though I only went there for five months.

Sharon honored Bud's successes with the Navy and shared that Bud was Sailor of the Quarter. She included his college education, his career goals, and accomplishments. It was wonderful to have Bud recognized for the man he had become.

In-Memory Card

After the twins and River completed the obituary, they focused on the in-memory card. They decided to use the Sailor of the Quarter picture for the in-memory card too. Bud would not have liked a verse, so the twins found a poem for the inside of the card.

> *Do not stand at my grave and weep.*
> *I am not there. I do not sleep.*
> *I am a thousand winds that blow.*
> *I am the diamond glints on snow.*
> *I am the sunlight on ripened grain.*
> *I am the gentle autumn's rain.*
> *When you awaken in the morning's hush*
> *I am the swift uplifting rush*
> *Of quiet birds in circled flight.*
> *I am the soft stars that shine at night.*
> *Do not stand at my grave and cry.*
> *I am not there. I did not die.*
> ~ Mary Elizabeth Frye

Birds Appear

The line *of quiet birds in circled flight* struck me. Since the deaths of Bud and Jeffrey, I have seen birds together every day. Two birds appear when I go outside, or get out of my car, and they squawk and squawk at me.

"Hi guys, I love you and miss you," I say, and the birds squawk and fly away.

Lately, I have two Ravens that come by every day. I hear them outside. "Caw, caw. Caw, caw."

Florida Common Raven—Blackbird

This goes on and on until I take them peanuts to eat. Even my neighbors caught onto the birds coming to see me. They say, "Bud and Jeffrey are here to see you."

My psychic said, "Bud comes as a hawk. He flies fast like the jets he used to fix in the Navy."

Bud the Hawk

BUD

On the Angelversary of Bud's fourteenth death day, Kim and I were playing golf. This hawk landed on the tenth green flag. Kim took the photo as I walked up, talking to Bud the Hawk. He stayed until I putted my ball into the hole.

Hundreds of times, Bud the Hawk has sat on a streetlight or in the tree outside our home, and when I go outside, he swoops down. He flies about a foot above my head as he squawks at me. I absolutely love it. This reminds me of the *Top Gun* movie, when Maverick did a fly-by, buzzing the tower as the Officer in Charge spills his hot coffee all over himself, swearing. Bud and I loved that movie.

I watched for signs from my children. I believed they would let me know they loved me and missed me. They were forever in my heart.

The Next Day Came. The freaking Next Day would not stop coming.

26

FAMILY FIASCOS

The little grandchildren of Lynette were adorable, even though seeing them reminded me that I would never have grandchildren. The realization of this was another crushing moment in my life. The blows just kept on coming. The Universe threw them at me, punch after punch.

My parents were not the greatest parents, nor were they good grandparents. I learned when I had children that babies did not come with instructional manuals. However, I chose to let Bud and Jeffrey know that a parent always had choices to be the better person.

"I did the opposite of my parents in everything I did for you. Both of you are smart, handsome, and can be anything you want when you grow up," I told them.

I never received this kind of encouragement as a child. I wanted my children to always know love and kindness. I am not saying it was an easy job to raise two boys by myself, but together as *The Three Musketeers*, we made it through the good and the tough times.

And Then There Was Torrance

Torrance, the Marine Corps Sergeant, called me when Bud died, and Kim made me take the call. He had disowned me when I was twenty-nine years old for dating a woman, so it was odd that he called me at all. Wyatt must have called and told him about Bud's death.

I gave him the benefit of the doubt. Surely, he called to share his condolences, about my son Bud, which he did. "Kathy, I am deeply sorry for your loss. I cannot believe you lost another child."

Gracelessly, he just could not stop with saying he was sorry, offering simple condolences under the circumstances. No, he had to rub salt into the wound. "I know you've done a lot of crap in your life. You certainly have a cross to bear. Kathy, it looks like you have a double cross to bear now."

There was nothing I had to say to this man. In my pain, I rose above his ignorance and was infinitely polite. "Thank you, I appreciate your concern." I hung up.

"Are these people mentally ill?" I asked aloud to nobody.

What the hell had he even meant? Did Torrance think all parents deserved to lose their children, or just me? I never received an answer to that question, or any questions about my upbringing. When I looked back at my childhood and the hell I went through, I know I was lucky to even be able to love my sons.

Thankfully, I loved my children more than my own life. I would have given my life if either, or both, of my children could have their lives back. After the phone call from Torrance, there was nothing left for me to give to anyone. I had nothing more to say.

"Kim, I am not talking to anymore people, period," I said.

Melinda Arises

Next thing I knew, the phone rang, again. *Will this never end?*

"Kathy, it's Wyatt. I'm not making this decision for you," Kim said.

I was so tired of all the bullshit that went on in my family. Kim handed me the phone and walked away, which left me little choice but to talk to him.

"Kathy, I am deeply sorry for what happened. Bud was a great young man."

"Thanks. Yes, he is. It's another useless, horrible loss!"

"Hey, listen, Melinda wants to come to Bud's funeral. Is that okay?"

I had not seen nor talked to Melinda and had no intention of ever talking to her again. The night I found out somebody murdered Jeffrey, when I called Melinda to tell her, I not only lost my son, but I also lost my mother. Unfortunately, life has a way of changing us whether we want change or not. Nothing really mattered anymore. My life, destroyed once again by the loss of Bud, was a disaster.

I knew it only hurt me to hold on to my anger against Melinda, as she would never admit that she had done anything wrong. So, in the middle of the obliteration of my life, once again, I took the high road and became the better person. "She can come under two conditions. First, she does not talk to me. Second, she does not say anything stupid."

Wyatt agreed, "Okay, thanks. Diana and I will bring her with us."

"Great."

"We will be there on Monday afternoon."

"You can stay at our neighbors, the Jerrys, across the driveway. They have two guest rooms."

Nothing would ever bring my two sons back. Nobody mattered to me anymore. Life was so unfair.

It was mean, and it was horrendous to me. Those were the facts of my life.

In the middle of an unbelievable, horrific, disaster, I forgave Melinda for all the things she had ever done to me; that list was long. I have no idea where that strength came from, but it came from a higher force than me. It could have come from Bud and Jeffrey up in heaven—who knew for sure?

The Next Day Came. The freaking damn thing just came.

27

COINCIDENCES

Can we just go back to the days when my boys were young and alive? Hurting both physically and mentally, I was running a negative balance on caring about anything or anyone. Tapped out and defeated, I gave up. What was the point of carrying on? I had no idea why I was still breathing.

Theatrical Events

Incredibly, Kim and I became aware of yet another coincidence between Bud and Jeffrey's deaths. The week after Jeffrey's death, Kim and I had tickets to the play *Mama Mia* in Poison Oak, California. Yes, Kim and I went to the show; she thought it would be a good distraction. I cried through the whole thing. It was not a show for a mother who had just lost a child.

Shockingly, two days after Bud's death, we had tickets to Cher's *Living Proof Farewell Tour* in Jumuba, California. Once again, we spent $150 each on tickets to an event. Kim felt we should attend the event come hell or high water—or in this case, the death of a second child.

I did not want to go to this event any more than I did *Mama*

Mia, but Kim believed in her heart that it would be a good distraction from the upcoming funeral services and the nightmare of my life. We went with a group of our friends, who all had a wonderful time; I cried. The only thing I do remember was Cher changed clothes for every song.

The most ironic thing about the concert was that one of Cher's greatest hits was the song "If I Could Turn Back Time." Unfortunately, life did not work that way. If I could turn back time, I would go back to when my sons were alive. I have since learned that time does not heal all wounds—it's how I use my time that matters.

The Blessing of Kim

The majority of what happened was a blur to me. Mercifully, when Kim called her work and explained what happened, they gave her time off to take care of things and me. Kim managed everything like a well-oiled machine—or was that organized chaos?

Kim focused on keeping everybody fed, showered, and where they needed to be. People brought food by the house, which kept everybody fed. I truly doubt I would have made it through any of this without Kim.

Death was good for business in the small town of Frostbite. Businesses made big money from Bud's funeral. There were over thirty people that stayed on the mountain, and more stayed down the hill. Kim had people in motels and friends' homes. The restaurants fed people, and the stores had shoppers.

Bud's Navy friend and ex-roommate Noah flew in from Ohio and stayed with us. He slept on one of the couches. Wyatt, Diana, and Melinda arrived and stayed with the Jerrys across the driveway. Renea and her friend arrived from Turquoise, Arizona. They camped out on our deck. Chad arrived; he slept on another couch.

The Navy Again

We made it through the weekend. The house buzzed with activity. Kim decided I needed to get up. I still saw no point to it.

"You'd better get up and get ready for the day."

"Why?" I asked.

"Travis said they would be here by noon for more Navy stuff."

Why does the Navy have to come again? I did not want to hear another word about Bud being dead.

Period, final, never. "I don't know what else we need to go over. You paid more attention than I did. What else is there to discuss?"

"Travis said he would be back on Monday at noon. It's Monday."

I got up, showered, and ate something, as Kim fed me like clockwork. *Really, I asked myself, why do I even bother? It's not like I have anything to live for, do I?* I certainly was not in a good state of mind; somebody should have recommended psychiatric help for me, don't you think?

September 1, 2003, while we waited for the Navy to arrive, I had a moment of sanity or insanity, I was not sure which way it went. I realized I had not heard any news about Bud's homicide from the California Highway Patrol investigation. Even though I swore I would never use the phone again, I dialed the California Highway Patrol. I wanted to know who killed my son.

Why do I always have to play games with law enforcement to get answers? Why do I always have to call them?

Police Response

An officer answered and identified himself by name and rank. I identified myself by name and *mother*. I said, "I am Bud's mother, and I would like an update on the information regarding my son's case."

He was happy to give me the report. "According to the report, Bud was struck by a hit-and-run motorist on U.S. Highway 501. The accident occurred when the driver of the vehicle abruptly switched lanes and hit the victim's motorcycle, causing his motorcycle to swerve from side to side before overturning."

"Yes, I got that part."

"Bud was pronounced dead at Pleasant Valley Hospital after the 6:15 a.m. accident near Santa Rosa Road. No one else was injured in the accident."

"Okay."

"A motorist who witnessed the accident pursued the hit-and-run driver but could not keep up. The suspect was driving a red Ford Taurus with two male passengers. All three of the people in the vehicle were believed to be in their twenties."

I could tell he was reading a report word for word to me. At least he was telling me information.

"There is no doubt the driver knew he caused an accident. It is the law to stop at an accident scene. This driver chose not to do that. The witness was able to get a license plate number of the suspect's vehicle."

"Well, that was good news, right?" I asked.

"The officers tracked that license plate to the owner in Rancho Santiago, California. The driver faces felony charges of hit-and-run, along with manslaughter."

How ironic. I went to Rancho Santiago three times a week as a truck driver.

"Have they found the people yet?" I asked.

Then, he told me something I did not know. "The police went to the house of the registered owner in Rancho Santiago. An elderly lady told the officer she sold the car to a Hispanic man six months earlier. She gave them the name and address of the man she sold the car to."

"So, the man bought the car from this lady and did not register it?"

"Correct, the officer determined this man was an undocumented immigrant and had not registered the vehicle. He believed the other two men with him were undocumented, too."

"Wow, isn't that just great?"

"The police went to the address they were given. A lady

answered the door, claimed she did not speak English, but told the officers that nobody by that name lived there. The police left."

The officer took a deep breath and hesitated for a moment. "Later that evening, the officers went back to the house. When they arrived, nobody answered the door. Upon further investigation, they realized everything in the house was gone, even the toilet paper."

I tried to remain calm, but my head was pounding and about to implode. "My son was killed while serving his country. He recently returned from Operation Iraqi Freedom and was in preparation to go back to that same war to defend our country's freedom. My son, killed on American soil, by undocumented immigrants, deserved to have his case investigated properly. Don't you think the California Highway Patrol could have left a police officer to watch the house?"

"Hindsight is 20/20, they say. They are continuing to investigate," he said.

At this point, I was about to lose it, so I thanked the officer for the information. I had to calm down before the Navy arrived. I would talk to Travis about what I learned.

One more coincidence. Jeffrey's story of injustice repeated itself in the loss of Bud. Again, my son suffered the injustice of the justice system, just in a different state, a different department. A botched investigation resulted in the loss of the suspects. Instead of the questionable arrest of one of three *bad guys* in Jeffrey's murder, there was no arrest of any of the three possible *bad guys* for Bud's homicide. There was no justice or closure, not for the victim nor his family.

28

MIRACLES HAPPEN, OR NOT!

It was high noon as I waited for the Navy to arrive. After the call to the California Highway Patrol, I needed fresh air, so I decided to sit out on the deck on the swing. Abruptly, I heard a vehicle turn into our private street. I looked up and about fainted on the spot.

Bud's Chevy truck pulled into the driveway! *OMG, I knew it. They were wrong! They were all wrong—the police, the hospital, even the Navy. Bud is not dead at all.* I thought I might have a heart attack right there on the swing. I could not breathe, as suddenly, I gasped for air. "Kim," I screamed, "Bud is ALIVE!"

"Thank you, God!" Kim yelled, as she flew out the sliding glass door onto the porch.

I knew they were wrong. I could not move, in fact, I would not move. If I stayed in this very moment, Bud was still alive.

The door to the truck opened. It was not Bud. *Bam*—crushed again! I blinked my eyes. It was Don who got out of Bud's truck. *What the hell?* Bud was not here. He was not coming home. Bud was never coming home.

The Nightmare Continued

Had Kim not been standing there next to me and caught me, I would have fallen to the ground. Why can't I just die and end this misery? Why is the Universe so incredibly mean to me? What have I ever done to deserve all this pain and madness?

Wtf? I could not believe the Navy did not warn me they were bringing Bud's truck from the base. I know they did not bring it to hurt me, but a warning sure would have been helpful. Don had driven Bud's truck from the Navy base. Rayna rode with Don so that he was not alone in Bud's truck. Travis followed behind them in his Navy car.

Useless Apologies

When Travis saw my face, he realized quickly that he should have called and told us they were bringing Bud's truck and possessions. He apologized, "I am so sorry for not telling you we were bringing Bud's truck up to the mountain. I wanted you to have Bud's belongings with you."

There was nothing I could say.

"Travis, Kathy was sure that Bud was not dead," Kim said.

"I'm sorry. We should have told you. Sorry!"

Bud had been staying in the barracks during his time back from his first tour, so Travis and Rayna had all his personal belongings. It would be a project to go through everything. There was no way I could do anything with his stuff at that moment. It was beyond the scope of what I could mentally manage.

Quickly, Kim realized it would be a smart idea to take advantage of all the people available and unload Bud's truck. Together, we emptied the truck and put all Bud's belongings in the downstairs room at the bottom of the spiral staircase. Kim and I would go through it later—much later if I had anything to say about it.

Don's Response

The world had changed so dramatically in such a brief time. It was just Sunday that both Bud and Don had helped fix the fence, shared dinner, and told stories. It had been such a wonderful day. To see Don again was both comforting and heartbreaking at the same time. Now, we shared the loss of Bud.

Bud helped Don not only in the Navy, but in his personal life. Although it made me cry, it was good to see Don.

"I am so deeply sorry, Kathy. I cannot believe Bud is gone. He loved you so much. I wanted to bring Bud's stuff to you personally, so no stranger touched his stuff."

"Thanks, Don, I appreciate it."

"Bud was my best friend. I don't know what to do anymore."

We hugged and cried.

Technical Navy Stuff

We went upstairs and sat at the dining room table again. Once again, Travis got out his big notebooks and spread them across the table.

"Again Kathy, I am truly sorry that we caused you such a scare," Travis said.

"I was so hopeful Travis; I want Bud back."

"This has got to be so hard for you," Travis said.

"Travis, you have no freaking idea. Before we start, I have some information I want to share."

I told him what the California Highway Patrol told me.

"The witness to the accident was Bud's commanding officer, Commander Gavin Sawyer. You will meet him at the funeral services on Wednesday," Travis said.

"Wow, I did not know that."

"The witness who got the license plate number was Captain Jones, an Air Force pilot, who flew with Bud's Navy unit. You will meet him at the funeral services also."

These two officers returned from war only to witness one of

their own killed on American soil. How extremely sad. Now, I realize how unfair this was to these brave folks who had put their lives on the line for this country and our freedom, only to see their comrade's life taken by people in that country illegally. Additionally, they did everything possible to catch those responsible for Bud's death, only to have their efforts wasted when the police botched the investigation.

Kim changed the subject and moved forward. "Travis, I take it you got the email I sent you?"

"Yes. Bud's obituary was in yesterday's paper."

"I also sent it to the *Fallen Meadow Times*, and *Bighorn Newspaper* in Utah. Liana put it in the *Tangerine County Registry*," Kim said.

"Do you have the information for the in-memory card?" Travis asked.

"Yes, the twins and River did a fabulous job on it," Kim said.

Travis looked at the photo on the cover. "Wow, I remember this picture being taken when Bud got the Sailor of the Quarter award. He deserved it."

"Bud loved the Navy," I agreed.

"Yes, he did. Kathy, today I have insurance forms and other documents that you need to sign as Bud's primary next of kin (PNOK) and person eligible to receive effects (PERE)."

I said, "Whatever."

We went through the necessary forms. I do not remember the number of forms or the purpose of each form. Believe me, I have no idea what I signed or did not sign. I was in shock, a fog, or whatever you want to call it.

My life had become a vortex of blankness. There were so many similarities between this and the shock and loss of memory when Jeffrey died. Sitting at the table with these people, there were no clear words.

Blah, blah, blah. Sign here. Blah, blah, blah. Sign there.

There are no memories of all these in my brain.

Funeral Details

Travis said, "The sailors from Bud's unit VR-55 all want to attend his funeral services on Wednesday. The Navy will supply buses for them to get to the cemetery before the hearse from the funeral home arrives."

That made me cry. It was touching to know Bud had affected so many lives. "That is wonderful for them to honor Bud. He will appreciate it," I said.

"Yes, they are adamant about attending," Travis said. "The question is what you would like them to wear?" Travis asked.

"What do you mean?" I asked.

"It's up to you what they wear. Do you want dress black or dress whites?"

"It is September and hotter than hell. They should wear their dress whites."

Travis laughed, "They will appreciate your choice."

I do not remember much else that happened at that meeting. Travis, Rayna, and Don wanted to head back to the base before rush hour in Poison Oak.

Travis concluded with Kim, "When you get to the cemetery on Wednesday, check in at the office. They will direct you to a sign with Bud's name on it. When you find the spot, park your car in line and wait. The Navy and Bud's commanding officer will meet you there."

We thanked them for all they had done and agreed to meet them at 1:00 p.m. at the cemetery. We walked them down to their car and said our goodbyes. Kim and I walked back up the spiral staircase from the driveway in silence.

Bud's eternal resting place would be at Riverside National Cemetery in Riverside, California. The twins researched and found out this was the third largest cemetery managed by the National Cemetery Administration, 740 acres donated in 1976.

"Bud wasn't even born in 1976," I mumbled.

I just wanted Bud to be alive. I needed Bud to come home. I was not going to make it on my own. There was no more *Three Musketeers*. *Two of a Kind* was gone. That left just me, *One*, alone. This was not how my life was supposed to turn out. Everything had gone wrong.

Bud completely prearranged his funeral before he went to war. He planned exactly where and how he wanted his funeral to take place. It was a true blessing to me as I could not have managed anything else during this time. The Navy and funeral home oversaw everything professionally. It was still the worst day of my life, a horrible day, but I did not have to make any decisions about the funeral or pay any money.

Helpful Information

Years later, I learned that at the time of the loss of a loved one, legal, financial, and estate decisions happen quickly. I wish I would have known to have legal representation when I made all those decisions. Nobody seemed to understand that I was incapable of making any decisions at that time or for years to come, yet they had me sign innumerable documents that had no meaning to me. The loss of my two children took me to places of complete desolation and depression, and I had no idea if I would ever return.

People might have thought I was there in the room with them at that dining room table, but I no longer understood the Universe as it existed. I remained at that table in physical form, but mentally I was a shadow of my past. With a broken heart and a crushed soul, all I could honestly think about was the demise of my two sons.

I believed the people who sat with me at that dining room table understood what I should or should not do. I put this accountability on the Navy and their CACO officer. I trusted that Travis knew what he was telling us and did his best at the time.

People who lose a loved one, especially during military service, be it a child, sibling, or a spouse, required professional guidance in understanding the ramifications of what they are signing and their rights. Even on a day when a sane person had not just experienced death and dying, the acronyms and forms would baffle a non-military person's level of understanding.

Kim and I dealt with legal issues involving the Veterans

Administration (VA) for years to come, resulting from decisions made and paperwork signed at that dining room table with the encouragement and guidance of the CACO Officer sent by the U.S. Navy. All those decisions made under extreme duress, loss, and confusion came with a price tag.

The military might consider including a legal representative in there CACO procedure for families during the signing of all those papers. I know Kim did her best to help me, but she had never been in a situation like the one we dealt with either.

All these people hanging around was a distraction from reality. The Next Day Came. No matter what, The Next *freaking* Day Came. They just kept coming.

29

THE LOVE RIDE (HONORING BUD)

Bud loved Harley-Davidson motorcycles and bought his motorcycle at Tropico Harley-Davidson. He spent all his free time there. No matter where Bud went, he made friends and people liked him. Bud even rode alongside *Jay Leno*, the late-night television star on two of his rides with the Harley-Davidson group. He was so excited when he got home and told that story.

Love Ride

Kim called Tropico Harley-Davidson and asked to speak with Gabe. "We wanted you to know the details of Bud's funeral in case anyone wanted to attend the services."

"Are you freaking kidding me? That was Bud on the U.S. 501 in that accident? The news called it a felony hit-and-run. Did they find the people yet?"

"No," Kim said.

"What the hell? I cannot freaking believe it was Bud. I can guarantee a lot of the riders will attend Bud's funeral services."

"Kathy will appreciate that. I know Bud will be honored."

Tropico Harley-Davidson, who sponsored The Love Ride for charity also honored Bud. They renamed the ride after Bud in 2003, the *Bud Love Ride*. They even accepted donations in Bud's name in lieu of flowers for the funeral. What a tribute to Bud.

Love Ride in Honor of Bud—2003

Later in 2003, Chad from Arizona rode in Bud's honor. He took this photo showing all the motorcycles that lined up to ride in memory of Bud and to support the *Bud Love Ride*.

The Next Day Came. The damn thing would not stop coming.

30

PAYBACK IS A BITCH

Artie and Gail arrived in California, from Wisconsin, on Monday September 1, 2003, two days before Bud's funeral service. Artie had not spoken to Bud or me since the child support incident. Amazingly, they arrived for another funeral of a child they ignored.

However, Artie and Gail did not come directly to our house. According to their communication with Kim, Artie drove to the scene of the accident on the U.S. 501 freeway. He told Kim that he wanted to see the exact spot where Bud died. Who knows why, but he told Kim that he felt he lacked all the answers about Bud's death, so he tried to find his own answers.

Police Questions

Artie's next stop was the nearest California Highway Patrol for the area. He introduced himself as Bud's father. Since he and Bud have the exact same names, other than III and IV, the officer assumed he was Bud's father and told him the same story they told me on the phone, nothing more.

Navy Stories

Artie's third stop was NAS Point Mugu in Oxnard, California. He wanted more information from the Navy about Bud. When Artie got to the base, the military police stopped him at the entrance to the base. He asked for Bud's unit, VR-55 Minutemen.

The guard directed Artie to park his vehicle and go into the visitor center for the information he wanted. They gave him the commander's name, and the phone number for VR-55. Artie dialed the number, and a man answered the phone. He identified himself by unit, rank, and name. Artie never shared what he learned at the Navy base with Kim or me.

After the Navy visit, Artie and Gail drove up to Frostbite and joined the group waiting for Bud's funeral at our house. Artie did not mention anything to me about his investigations or what he found out. They stayed for a while, and finally, Kim took them into Frostbite to their motel. Kim was much kinder to Artie and Gail than I would have been. Personally, I had nothing to say to them.

Waiting, Alcohol, and Bud Stories

There was nothing to do but wait. There were dozens of people at our house. I do not remember who was there. Unfortunately, I do not remember all the stories, but I remember the feeling of love these people shared for Bud. There were stories from the Navy and stories from Bud's childhood. The more alcohol consumed; the more stories shared about Bud.

Countless of these stories were not meant for a mother's ears to hear.

NCIS Television Show

One of my favorite stories about Bud was a Navy story. Most people have heard of the television show *NCIS*. They filmed scenes that included Navy personnel working on airplanes, or the team boarding Navy planes, at Bud's unit at NAS Point Mugu.

Bud came home from work one day upset.

"Bud, what's going on?" I asked.

"They are filming a television show at our hangar."

"Wow, that's cool. Why is that a problem?"

"We have airplanes to fix, Mom; we don't have time for TV shows."

"It could be fun."

"They asked Don and me if we wanted to be extras on the show."

"Bud, you said yes, right?"

"No Mom, I told them we didn't have time."

"Well, it might have been fun to see you on television."

"Mom, this short gray-haired guy came over and told us we were making too much noise."

"You mean *Mark Harmon*?"

"Who's that?"

"Remember that movie you and Jeffrey watched, *Summer School*?"

"Are you kidding, the teacher?"

"Bud, you are too funny."

"I didn't know it was him. He is short."

"Do you think maybe you are just tall, Bud, at six-three?"

Bud always made me laugh.

Study for the Navy

Another one of my favorite Bud stories took place in 1996. He bought a book that prepared him for the Armed Services Vocational Aptitude Battery (ASVAB) test. One day, as Bud and I drove to Fallen Meadow, Utah, from Turquoise, Arizona, to visit Melinda, Bud read the questions in the book.

Neither one of us paid attention to my driving. Traffic was light, and the road was straight and boring. Next thing I knew, there was a police car next to me. Frustrated, the officer honked his horn and waved his arm and hand for me to pull over. His sirens blared and lights flashed.

"Mom, you are toast. You're getting a ticket." Bud laughed as I pulled over.

The officer came to the window of the car. "Ma'am, do you know how fast you were going?"

"No officer, I do not. My son is studying for the ASVAB test, so he can join the Navy. We were just driving along answering the test questions."

"Well, you were doing 80 mph in a 55-mph zone."

"No way, I never speed. I have driven over a million miles without a ticket or accident."

"Ma'am, driver's license and insurance."

I got out my wallet for my driver's license, and Bud got out the insurance from the glove compartment.

"Officer, I drive for Box Trucking in Turquoise. I work with over a hundred guys. I am the only woman driver. If I get a ticket, I will never hear the end of it."

That did not seem to matter to him either. He took my information. "Thank you, Ma'am."

As he walked away, Bud laughed and laughed some more. "You're getting a ticket. It's going to cost you big money."

The Officer came back to the car. "Here is your license and insurance. Sign here."

I was not happy, but I was polite as the officer smiled at me. "Have a wonderful day, Ma'am. Young man pay attention to your mother and study hard. Make her proud. Ma'am again, have an enjoyable day."

I thanked him as he walked away. I looked down at the ticket. It was a non-moving violation ticket for wasting natural resources. It cost me ten dollars and would not go on my record.

Bud was shocked; he could not believe it. "If I were driving, I'd be going to jail. You shed a tear and tell your sad story, and you get a ten-buck ticket. Must be nice to be a girl."

"No Bud, I was respectful. I told the truth. I explained the situation calmly. He went back to his car and ran my information. He saw I told him the truth because I do not have tickets. He was a decent person and helped me out. He was impressed you were studying to make yourself a better person."

Bud looked at me and thought about it. "You know Mom, I think you're right. You saved yourself a ticket, and he saved his dignity. Wow!"

"Yes Bud. Remember, I used to be a law enforcement officer, too."

It was particularly important to teach Bud things that he needed to know as an adult in our society.

As a parent, it was my job to lead by example and explain when necessary. As a single parent, I had to bear double the responsibility of not only ensuring my sons grew into functioning adults, but that they were respectful males in society.

This incident had a huge effect on Bud's character and the way he looked at and lived his life. It taught Bud how important it was to have integrity and to be honest. He saw what it meant to remain calm and be a good person who told the truth. He also learned the police were not the enemy if you were fair and respectful with them.

Skiing with Renea

I shared one other story that involved both Bud and Jeffrey. I met Renea and her family during my divorce from G.I. Ranger. Renea did not like to hang out with my kids; nonetheless, when we went snow skiing three or four times every winter, we always took Bud and Jeffrey along to the ski resorts in Southern Utah. Renea paid for them to go to ski school every time. I dropped them off at 7:30 a.m. and picked them up at 5:00 p.m. This went on for three or four years.

Eventually, Renea and I parted ways, even though we are still friends today. A year later, the boys and I decided to go skiing. It shocked me how well they could ski. They went down the black diamond runs and jumped anything that came in front of them. Bud loved to snowboard, and Jeffrey loved to ski.

"Holy crap, you guys can ski!"

They looked at me like I had lost my mind. Bud said, "Mom, I think your hair dye has soaked into your brain. We went to ski school. What do you think we were doing all those times you dropped us off? We learned how to ski."

"I don't know. I thought you just played inside. I knew you got breakfast and lunch."

"I guess Renea got her money's worth for those ski school days," Jeffrey said.

"Yes, she did," I agreed. "You two are great skiers."

Everybody loved my stories.

Thoughts

If you took the horror of Bud's death out of the equation, all the stories told were funny and full of love. We all laughed about it. Humor was better than crying, and all the alcohol consumed helped the stories flow. I wish I could have had alcohol, but Bud and Jeffrey would have been disappointed, so I did not have any.

Sarcastic humor got me through unbelievably horrible stuff for years to come. There was no way to rationalize all the terrible things that happened in my life, so I found ways to disassociate myself from it. Humor was a release—not funny *ha-ha* humor—sarcastic dark humor.

It had been a long day, and the next day would be a horrible day, no doubt about it. I missed Bud so much. The *Three Musketeers* shared the good and the bad. We worked, laughed, and cried together. Best of all, we loved each other.

Bud had so many things he wanted to do. All his hopes and dreams were gone. So much potential wasted. Reality was about to hit me square in the face. Left in the world without my children, I thanked God that Kim came into my life before *They* took my sons. Without Kim, I would never have made it through these horrible events in my life. I was tired.

The Next Day Came. Unfortunately, this time the damn Next Day came.

31

FUNERALS SUCK

The most dreadful Next Day Came. The day a parent never imagined or planned for arrived on my doorstep. *How can I make it through this day when I cannot even breathe?* There was no air to breathe. I had already buried one child. I wondered what more I owed this cruel world. What more would I have to pay for whatever atrocities I might have committed?

No parent ever envisions they will lose their child. Parents should never have to bury their children. The Universe had flip-flopped in my life. Everything was upside down. I had to be in another dimension because this one sucked.

Declaration

This would be a long, horrible day. When my children were alive, I saw the world in color. Now my world had turned black and white. There was no power left to draw on for strength. I had not slept, but when Kim awoke, we shared a discussion.

"Kim, I do not want to do this today or any day. What will I do without Bud? He has been my friend for twenty-four years. It was not enough time with him," I cried. I knew in my heart that Bud and Jeffrey were together in heaven, but I wanted to be with them. I did not want to be the one left behind.

"Kathy, I know, but it is time to get ready for the funeral. As Bud's mother, you have important things to do today. You must stand for Bud with pride and dignity. There will be many people there to say their goodbyes to Bud."

"I declare right now that I do not want to do this. Do you get it?" I demanded.

"Yes, I do, Kathy. I will do whatever I can to help you through the day. We will do it together. We will make Bud proud, okay?"

"I will do this for Bud because I love him."

Kim had taken me shopping over the weekend for what we called a funeral suit. I despised shopping. Thankfully, we found a black three-piece pant suit I would wear. I put the suit on, we climbed into the truck, and Kim drove down the mountain to the Riverside National Cemetery.

Peggy's Story

Kim had planned for a gathering at our house after the service at the cemetery to celebrate Bud's life. We had no idea how long the funeral service would take, nor the number of people who would attend. Supporting the mountain community, Kim ordered food at the local grocery store in Frostbite. Peggy and Sue volunteered to stay at our house during the funeral to set up the food and drinks. We no sooner reached the bottom of the mountain when Kim's cell phone rang.

"Kim, I brought deviled eggs. Do you have any paprika?" Peggy asked. "I've looked everywhere for it."

"If we have any, it would be in the pantry. The big cabinet by the stove top."

"I've already looked there."

"I guess we never use it, Peggy."

"Kim, I can't believe you never use paprika. You are the one who cooks in this group. Never mind, I'll make do." Peggy hung up.

"Kim, that phone call had nothing to do with paprika. Peggy doesn't even cook," I said.

"I know, Kathy. No one knows what to do or say about today. Bud should not be dead."

Riverside National Cemetery

It would be a long day. Everybody's nerves were on edge. For those who knew Bud, it was going to be a horrible day whether they attended his service or not. Nobody wanted to attend a funeral, much less one for a twenty-four-year-old young man whose life ended senselessly.

As Kim and I drove into the cemetery, overwhelmed with sadness, and swept with awe at the reverence of the property, we stopped at the cemetery office. It was a long drive down from the mountain, so we asked for a restroom. The lady at the desk pointed us in the right direction.

"Okay Kathy, just breathe."

"I think that's about all I can do right now, Kim."

When we returned from the restroom, the lady inquired if we needed any information. "Is there anything I can do to help you? Do you need directions?"

Kim said, "We are here for Kathy's son's funeral."

"I am heartbroken that another young man has lost his life. You have my condolences." She directed us to the area we needed to go, and we thanked her for the information.

A Mother's Nightmare

Kim and I drove to the designated parking area. As we pulled up, we saw Bud's name, rank, and U.S. Navy on the sign, just as Travis told us we would. We parked our car at the front of the line and waited, but not for long. With shattered nerves, we got out of the car and walked up and down the sidewalk.

The Navy arrived in two buses and two vans. The vehicles pulled in along the curb behind us and parked, and I was surprised there were so many people. The bus doors opened, and a sea of white uniforms flowed out. This site was like watching the white caps of the waves on the ocean crash into the beach. I had a new understanding of the term *sea foam*.

It was an awe-inspiring sight to say the least—incredibly sad and magnificent at the same time. They lined up on the sidewalk down from us. I know Bud would have been so proud that all these Navy personnel showed up to honor and respect him.

Kim and I stood there in amazement and watched as the sailors lined up and prepared to honor their fellow sailor. Kim lacked any experience with the military. She knew Bud, Noah, and Wyatt, but she had never attended any military events. She admired the precise assembly and professional attitude these young service members displayed.

I loved the military, their protocol, and precision. When we lived in Hawaii, G.I. Ranger was the most decorated enlisted person of all five military branches. Since Bud was in school, Jeffrey and I spent three years going to every military event imaginable with G.I. Ranger, including change of commands and retirements. I tried to find G.I. Ranger or his children to tell them about Bud's death, but I was not able to find any of them. I have no idea what happened to them after the divorce.

Kim and I continued to watch the sailors line up and prepare for the service.

The Navy Arrived for Bud's Service

Two officers, both in the U.S. Navy Reserves like Bud, walked up to us and introduced themselves by rank and name. The first officer was Commander Sawyer, Bud's commanding officer. Kim thought the commander looked like Robert Redford. She referred to him as such for the rest of the day.

The other officer was in the Air Force, Captain Jones. He was the one who chased down the car that hit Bud. He got the license plate number and gave it to the California Highway Patrol. We introduced ourselves to them. Kim, always good at these kinds of things took over and started to ask questions. "What do you do when you are not at the Navy Reserves?"

Commander Sawyer answered, "I fly commercial 747s for Delta Airlines."

"I also fly for commercial airlines," Captain Jones said.

"Maybe I've been on one of your flights; I fly a lot for my work."

Everyone laughed, and it broke the stress a little for this horrible situation. I had nothing to say, so Kim asked more questions. "You both have a lot of medals. Are they like swim meet medals? Bud used to swim competitively, so did his brother Jeffrey. They have a lot of medals and ribbons."

The officers got a great chuckle out of this. "Yes, they are similar to medals won at swim meets," Commander Sawyer answered.

"How long have you both served in the military?" Kim asked.

I could tell they were trying to keep things light. The men spoke kind words about Bud. It was heartwarming for me to know the lives Bud affected. They remembered something and became excited.

"NAS Point Mugu has a charity golf tournament later in September. In honor of Bud, we renamed it the Bud Golf Classic."

"What an honor for Bud. He loved to play golf with Kim and me. After Jeffrey's death, we hit buckets of balls at the driving range. At first, Bud was so angry that he broke one of my drivers."

"Yes, I remember Bud telling me that story," the commander replied.

"Bud and I found it a terrific way to release frustration. We smacked the crap out of those little white balls."

"I agree. We would like you and Kim to join us for the tournament. Would you be interested?"

"Yes! We would not miss it. Just let us know where and when. We will always be there, for anything that honors Bud. Thank you."

Commander Sawyer said, "I also should tell you about Bud's father, Artie, calling me. I'm not sure you were aware of him coming to the base, but things did not go as expected for Artie when he called from the entrance gate to speak with me about Bud's death. First off, let me say, it was shocking to hear how much Bud and his father sounded alike."

"Yes, they sound and look alike. He is here for some unknown reason."

"Well, according to the U.S. Navy, Bud's father is deceased."

Shocked, I laughed aloud. I looked up at the sky and smiled. "Way to go, Bud! Way to go!"

"Yes," Commander Sawyer said. "Artie was quite shocked to learn that information."

"Oh, that is precious," I replied. "No wonder he did not tell us what happened."

"It appears that when Bud joined the Navy, he told the MEPS people his father was dead. When Artie called, I told him: *Since your son listed you as deceased on his entrance forms, I am not at liberty to give you any information.*"

"That is hilarious," I said.

"Yes, Artie was not happy with my reply at all and hung up the phone."

Bud got a good stab back at his father for all the years of hurt and heartache he caused Bud, Jeffrey, and me. Truthfully, at that point in my life, it did not matter anymore. Hard to believe, but for the briefest of moments, thanks to Bud, I laughed and smiled at his funeral.

Funeral Service Description

Commander Sawyer grew intensely serious. He felt he needed to describe the procedure for the funeral ceremony. He wanted us to feel as comfortable as possible with the procedures and know exactly what would happen during the service. I would never feel comfortable, but we listened.

It is time to put on my mom face and get through this horrible day.

"First, when the funeral coach arrives, all military members will stand at attention and salute."

"Okay."

"The pallbearers will take the casket out of the funeral coach."

I almost fainted at that statement but managed to stay upright. That was it; with the mention of a casket, I blocked everything else out.

"The chaplain will lead the way to the gazebo, and the pallbearers will place the casket and secure the flag."

Kim took over for me. This pattern continues, even today, when I do not want to deal with situations. Diagnosed with post-traumatic stress disorder (PTSD) and complicated grief, when the situation overwhelms me, I check out.

"So where will Kathy be during this time?" Kim asked.

"The family will follow the casket to the gazebo, and when the casket and flag are secured, they will take a seat. Don't worry, we will let you know when to sit."

"Thank you."

"The chaplain will perform the service. When he finishes the service, all family members will stand for honors."

"Okay," Kim said.

"Then the twenty-one-gun salute will happen. Let me warn you, this will be loud. It could shock you."

"Kathy will be okay with gunshots. I'm not sure about the rest of us, but we'll make it through."

"When the twenty-one-gun salute finishes, the bugles will play taps."

"That will be a very emotional ceremony."

"After taps, the Navy personnel will fold the flag on top of the casket, then I will present the flag to Kathy."

I found my voice. "Thank you."

"It will not be a long ceremony."

"That is probably a good thing, for Kathy's sake."

"I just wanted you to know it will be dignified and honorable for Bud, for Kathy, and your family."

It was good to know what would happen during the ceremony, but nothing really registered in my brain.

"One last thing, after the funeral ceremony is over, we will form a reception line. Don't worry, we'll help you with this, and I will stand by Kathy to greet the people and introduce them to her."

"Great, that will be very helpful."

Compliments for Bud

"Everybody will want to come by and talk with Kathy about Bud. He meant a lot to them, and they want to convey their condolences."

Kim said, "Thank you. I know Bud had the utmost respect for the Navy, and it's good to know they had the utmost respect for him too."

"We were all honored to know Bud. He was an exceptional young man."

"Thank you, that means a lot to me that you knew how special he was," I said.

"Believe me. He changed the whole attitude and atmosphere of VR-55."

"Bud was a great young man. He had so many things he wanted to do. This is incredibly wrong."

The commander hugged me for a long moment. "I absolutely agree, Kathy. This is all so incredibly wrong!"

I never imagined I would attend a funeral for my child, yet here I was at my second funeral in two years. I saw military funerals with honors on television or in movies, but I never imagined I would be at one for my son. If anyone deserved to be honored and respected, Bud certainly did. He served his country with honor and dignity. He so loved the Navy and life itself.

Harley-Davidson Motorcycles

As we finished talking with the commander, I heard a low and deep rumble. I realized it was the sound of Harley-Davidson motorcycles in the cemetery. I knew Bud's friends from Tropico Harley-Davidson had arrived.

There were more than fifty motorcycles. They pulled their assorted styles of motorcycles up to the curb and parked. The motorcycle group was quite a contrast to the over one hundred sailors lined up in their dress white uniforms.

I imagined Bud smiling in heaven. His two favorite worlds collided in one place for him. Albeit at a cemetery, the Navy and his favorite Harley-Davidson motorcycles were together to honor and bid farewell to him.

Apologies

Wyatt was next to arrive, dressed in his military uniform. He retired after serving thirty years. He walked up and hugged me. "Can we talk privately for a moment?" We walked down the sidewalk. "Kathy, I respected Bud and was proud of all he had done. I am so deeply sorry for all the things that happened."

"Thanks."

"I am so sorry for how I treated you when Jeffrey was murdered. It had to be horrible. I should have called you. I am sorry."

"Thank you."

"Thank you for asking me to be in Bud's service. You must be in unbelievable pain. If there is anything, I can do for you, please let me know."

"Thanks, Wyatt."

My life was one freaking nightmare after the other. The treatment I had received after learning of Jeffrey's murder, other than Lynette and her family, still stung. One would think I'd have immunity after all the years. My family continued to simply think I should be the one to just forget, forgive, and move on.

Somehow, at forty-three years of age, at another funeral for the loss of my only child, I decided these people would never hurt me again, period. What they did or did not do, no longer had meaning to me. Nothing had meaning anymore.

Bud's Arrival

The black hearse arrived. I knew Bud was in that vehicle. When the pallbearers took the casket out of the funeral coach, reality hit me, as if an inhuman force ripped the skin off my body. The pain was excruciating and indescribable. *Bud is in that casket, dead. He is never coming home!*

Reality Hit as Bud Arrived

Thankfully, Kim caught me before I fell. Shock hit me like a two-by-four in the gut. Déjà vu! The exact same thing happened two years before when I saw Jeffrey wrapped in sheets at the funeral home. Although I did not physically see Bud, I knew in my mind's eye that Bud was in that horrible wooden box wrapped in an American flag. It was just like seeing Jeffrey in those white sheets.

I knew my life was over when I saw Bud's casket. I would never see Bud again. The validation of the moment overwhelmed me. Somehow, I managed to remain standing and did not pass out. In my mind, three thoughts went around and around like a merry-go-round.

If only I could die. They could bury me along with Bud. There is enough room in the casket for me.

Out of respect for Bud, there were six pallbearers who requested to carry Bud's casket. Wyatt was the rear pallbearer on the back-right side. Wyatt said later, after the service. "When I carried Bud's casket, it was the hardest thing I have ever done in my life. It broke my heart, Kathy."

The middle pallbearer on the right side was Chad, and the front pallbearer was Noah, Bud's longtime Navy friend. On the Navy side, the front left pallbearer was Bud's best friend, Don. The middle pallbearer was Rayna, and I do not remember who the rear pallbearer was.

Suddenly, I heard Melinda's voice in my ear. "Kathy, you look as white as a ghost," she said.

If only Melinda, if only! I said nothing, nor did I acknowledge her presence.

My Memories of Bud's Service

I assumed things happened exactly as the commander told us they would. I have few recollections.

- The loudness and unexpected explosions of the guns startled me; my brain recorded it.
- The bugles played taps; I cried. To this day, whenever I hear taps, I cry.
- I received the flag from the commanding officer. The men folded the flag with the twenty-one shells fired during the twenty-one-gun salute. Kim and I learned this later, when we put the flag into a beautiful box, along with Bud's medals and military paraphernalia that sits on a shelf in the master bedroom. Next to the box stands a replica model of a F-18 Fighter Jet and a C-130 cargo plane, the latter given to me by the sailors of Bud's squadron.
- The chaplain spoke, although I have no memory of this. I have a copy of his sermon.

- I stood in the reception line; held up by the commanding officer. I collapsed into tears repeatedly. There were over two hundred people at Bud's funeral service. Condolences and comments shared about Bud were heartfelt.

Other Attendees

In the reception line, I recognized a girl that I thought Bud had dated. Her hair was dark before, but her hair was blonde that day. Amelia drove ten hours from Fallen Meadow and back. She never said a word to me or Kim. Bud must have influenced her life for this reaction.

Marisol attended. Later, she told us that she followed the casket. "I did not care who thought they was Bud's girlfriend, current or past. Bud was my best friend. I was part of this family, so I sat by you."

Marisol will forever be part of my family. She was surprised to see Noah. Marisol also introduced us to Bud's former boss, Gabriella, from Poison Oak Mission College.

Gabriella said, "Bud advanced the veterans program at Mission College. He loved helping other veterans and got them the benefits they earned and deserved."

Bud's Military Headstone

The service was a *big fat, mean, terrible, inexplicable thing.*

32

LYNETTE'S RECOLLECTIONS

After Bud's service, I noticed Artie standing by himself at the cemetery. In total disbelief, Artie and his wife Gail attended another one of my nephew's funerals. How could they? Gail was the very woman that Kathy caught dancing with Artie on that fateful night. That story stuck with me all these years—she was eight months pregnant with Jeffrey and her marriage destroyed.

I had flown to Wisconsin and helped Kathy drive those little boys to Utah when Jeffrey was still nursing. The story had pissed me off then, but I felt the slap in the face of them showing up at Jeffrey's funeral even more. No words described the anger and disgust I felt at Bud's funeral.

Blown away by the way that Artie treated Bud over the child support payment incident, Artie never stopped twisting the knife in my sister's back.

It truly bugged me that Artie had never taken any interest in Bud's and Jeffrey's lives, other than their deaths. Why would he bother to come to both funerals? It was clear to me that I would never see Artie and Gail again, as Kathy had no other children. So,

I figured: *Why not ask Artie the question that certainly had to be swirling in everyone's mind?*

"Artie, I have to ask, why did you bother to show up at both funerals for Bud and Jeffrey, when you never even bothered to get to know them during their lives?"

He looked at me blankly for a moment. Was he trying to figure out why he was there even for himself? "Lynette, I always thought I'd get to know them when they were older," Artie said.

Funny, it was my turn to look at him flabbergasted. Really? That was his answer? Suddenly, I saw things clearly. Bud and Jeffrey had been better off without him in their lives all along. "Well Artie, I guess you missed that opportunity because Bud and Jeffrey will never get older."

Artie stood there and stared at me; he said nothing. I could not believe what he had said to me. I turned and walked away, shaking my head in disbelief. I hoped he lived a long life spent questioning the losses he had. He would never know what he missed. I got in my car and drove back up the mountain to Kim and Kathy's house.

When I arrived at their house, I took them aside, and shared my story with them.

"Lynette, you're right. Artie missed the opportunity of a lifetime. He will never know who Bud and Jeffrey really were. Mercifully, he and I produced no other children, so this will be the last time we will ever see them," Kathy said.

33

ENDURING STORIES OF LOVE

Not ready to deal with people, I went out onto the porch and sat in the swing. The day was a blur, but after hearing Lynette's story, I thought about Torrance's words to me on the phone. If he believed that I had a cross to bear, then Artie had a huge cross to bear—a double cross.

It my heart and mind, Artie lived in an unimaginable hell, swimming in his guilt. It impressed me that Lynette had the nerve to ask Artie her question. I never would have bothered to ask him why he came; his reasoning made no difference to me. Neither he nor Gail made any difference to me. I would never see them again.

Thoughts and Memories

Artie and I produced two beautiful sons, but Artie decided he did not want a family. That day Bud, Jeffrey, and I left Wisconsin in 1983 was a huge, scary step for me. I was twenty-three, Bud was three, and Jeffrey was eight months old. I had never lived alone or even had a real job. Together, we not only survived—we grew up, and we made it through. Now, they were gone; people had senselessly killed them both. I was alone, the *One* left behind.

Their father never knew the relationship that Bud, Jeffrey, and I shared. Artie missed the love, the humor, the joy, the excitement, and the silliness these two boys shared in my life. Artie missed their friendship, their unbelievable intelligence, and the understanding they shared for the world and the people in their lives.

Lynette's story brought up a memory of something that Bud told me after Jeffrey died. "Mom, just so you know, Jeff and I were always just a little bit scared of you. We knew that we would never get away with any crap with you. Somehow, you always knew what we were up to."

"Bud, I always knew what you were up to because I did all the things you two thought of. Well, except the tattoo thing Jeffrey did. I would never have thought of that one. I am confident that tattoos, at fourteen, would have been right up there on his list of things for Torrance to kill me over. He threatened to kill me if I came home pregnant in high school."

Bud and Jeffrey were two of the most wonderful people I have ever known. They were my best friends, the loves of my life. They respected me, and we loved each other.

The rest of the evening consisted of large consumptions of food and alcohol, stories, and outflows of love for Bud. It was an incredibly sad and unbelievable event that I will never forget. This was something I never wanted to be a part of, but to see the lives Bud had genuinely affected was heartwarming to me. The people who came to our home after the funeral had the utmost love and respect for Bud. They would certainly miss Bud very much.

There were all sorts of Bud stories to go around. The stories they told about Bud revealed how he altered their lives and the lives of others. Bud had changed boys into men and even transformed sailors' careers. I was so immensely proud of him., I was sure Bud never thought his mother would hear the stories I heard from his Navy friends, but everyone laughed and cherished knowing Bud. They shared so much joy.

BUD

Explanations

The young man who answered the phone calls at Bud's Navy Unit VR-55 came to our home after the service. His name was Sam. After three or four drinks, Sam became chatty and shared the following stories with the group.

The first story was the phone call Artie made, and the second story was the phone call I made on the day Bud died. Unfortunately for Sam, he was the one who answered both phone calls.

The Story of Artie's Call

As told by Sam

"When Bud's father called, I answered. The voice on the phone asked to speak to Commander Sawyer. I about shit my pants. Sorry, ma'am. I meant that I almost dropped the phone. The person on the phone sounded exactly like Bud. In fact, I thought for sure it was Bud—from hell or the spirit world and he called, just to f**k with me. Sorry, ma'am," Sam said.

I laughed. "Yes, Bud and his father sound and look alike."

"Yeah, I know. I saw Artie at the service. He looks like an older Bud. It was freaking scary how Bud and his father looked and sounded alike. Funny thing, Bud never talked about his father. In fact, from the few things he said, we thought he was dead."

"Yes, it truly is strange how much alike they were, since Bud did not grow up around him. According to Commander Sawyer, Bud told the Navy when he joined that his father was dead."

Kathy's Call

"I also answered the phone the day you called when Bud died. You said you were Bud's mother and asked to talk with him. I about shit my pants. Sorry, ma'am. I didn't know what to say. We already knew Bud was dead. They told me not to tell anything to anybody who called. I freaked out and hung up," Sam said.

"Yes, I know you hung up. I said, 'What the hell.' Then I called back, and nobody answered, so I figured you were busy getting ready to go back to war."

"Nope, we all stood there and stared at the phone. It was like looking at the demon seed from hell every time it rang. Nobody wanted to answer it. Nobody wanted to be the one to break your heart, ma'am."

"Sam, truthfully, if this were not about such a horrible event, the stories would be hilarious. Thanks for sharing them," I said.

"I just wanted you to know how sorry I am, ma'am. Bud was a great guy, and we will all miss him. He was funny and worked hard. We all liked him. I am so sorry, ma'am," Sam said.

Marisol Story

"Kathy, at one moment during the service, I stared at you and thought to myself: Kathy must be the strongest person I have ever known. How were you even standing and breathing at that point?"

"Marisol, I don't think I am strong. Right now, I just survive moment by moment, breath by breath."

Marisol reflected for a moment. "I was sitting here and looked around at all the people who came up the mountain to share the celebration of Bud's life. It became clear to me that I was not the only one who had lost Bud. So many people will miss him."

"You are right, Marisol. Bud had so many friends. These people love Bud. He was one of a kind."

"Kathy, I wanted to say something, but I couldn't think of anything that would help the situation, so I stayed quiet."

"That's okay, Marisol. It hurts to lose a child, but it helps to know other people remember and love Bud. Most even knew Jeffrey through Bud's stories."

"There are no words that are ever going to fix what you've lost," Marisol said.

"I know, Marisol. Believe me, I know."

"Kathy, what can anyone say to a mother who has lost two children in two years?"

"Marisol, saying anything is better than saying nothing. Just say something. Truthfully, just say anything!"

"I'll remember that Kathy, thanks."

"Marisol, the person who loses a child feels so alone. There are no words to describe the emptiness. When I lost Jeffrey, I thought I would die, but I did not. I had to help Bud. I still had a purpose. I have no idea what I am supposed to do now. I have no purpose."

"I wish I knew how to help you Kathy," Marisol said.

"I don't know how I am going to even get through the next moment. Never in my wildest nightmares did I think it could happen twice—but it can, and it did."

Exhausted, my brain shut down. I was so tired of crying. I had a horrific headache, so I took my migraine medicine and tried to rest. *Can anyone count the tears one person can shed?* I must have fallen asleep after tossing and turning for hours during the night.

The Next Day Came. When I opened my eyes, I wondered why I had to wake up at all.

34

TAPS PLAYED FOR BUD

Go to sleep, peaceful sleep, may the soldier or sailor, God keep.
On the land or the deep, safe in sleep.
—Horace Lorenzo Trim, "Taps"

My eyes opened. The Next Day insisted on coming whether I had any interest in it or not. I had no children, no plans, no life, nothing to do, and nowhere to go. My life had ended, but I was still alive. *Why?*

I was awake but saw no point in getting up, so I stayed in bed. Maybe if Kim thought I was asleep, she would not make me get up and eat something. *Damn this diabetes stuff! What an inconvenience.*

People all around the house started waking up. They showered, and Kim made breakfast for anyone who wanted it. The people who had come for Bud's funeral would head down the mountain and home soon.

"You'd better get up. You need to shower and dress. People will be leaving soon, and you need to say your goodbyes," Kim said.

"Whatever, I'll get up. What else do we have to do today?" I asked.

"Not much, we need to get people on their way home and get the house put back together. I told Wyatt we would go out to dinner tonight."

"Why would you do that?" I asked.

"They wanted to spend a little time with you before they head back to Utah. Wyatt said they would leave tomorrow."

I shrugged. "We can have dinner if you want. I have nothing else to do."

"Okay, good. Shower, and I'll fix you breakfast."

Busy Work

After everyone left to go down the mountain, back to life, work, school, or the Navy, Kim decided we needed to clean the house. This would give me something to do and keep my mind occupied for a little while. Luckily, Kim and I always worked well together.

Kim and I did not go down the spiral staircase or anywhere near Bud's belongings. That was sacred ground. That moment when hope filled my heart, when I saw Bud's truck, and thought Bud was alive, still burned in my memory.

The Navy placed what remained of Bud's life in the room downstairs, at the bottom of that spiral staircase. I was like Hogan, my standard poodle, who hated that spiral staircase. He would not go down those stairs, and I would not go down those stairs either. There were monsters down there.

I had no desire to look at, much less touch, feel, and smell what *They* had left me of Bud, my son. Bud's belongings could just stay down there, untouched.

The day passed quickly, evening came, and we went to dinner at Castaways, a rustic restaurant set on a mountain top, overlooking the valley. Castaways was one of Kim and my favorite restaurants. I looked around the table, remembering the pain and misery these people had put me and my children through. *Do I need family in my life? Maybe.* I sat at a table with other people, in a crowded room. I felt so alone, empty, and depleted. I had nothing left to give to anyone, past or present.

New Realities of Life

After everyone went home, the fragility of life became clear to me. Life guaranteed nothing other than death. Things I used to think were vital were not the least bit important any longer. There was nothing that compared to the loss of my two sons.

On March 28, 2001, I found out somebody murdered Jeffrey. That left me with a hole in my soul and a broken heart. My emotions flatlined. Highs and lows disappeared in my life; I was virtually non-emotional. I did not want to be alive, but at least I took part in Bud and Kim's lives—or at best, I was present.

Fast forward two years and five months to August 28, 2003, when I found out somebody killed Bud. Suddenly, there was nothing left of my already broken heart; it shattered into a thousand pieces. Like a lithotripsy and a kidney stone, or Humpty Dumpty's fall off a wall, there were only tiny pieces floating around in my chest. I knew that no one would ever be able to reassemble my heart.

The focus of my entire adult life had been the care of my two boys. Where was I supposed to find the strength to continue this time? I had Bud who needed my help the first time. Nobody needed my help now. I had no purpose left. Emotionally, I was empty, tired, and drained. I had nothing to live for. I wanted to see my children again.

Important Revelation

Looking back, the most decimating thing about Bud's death, which should have been the clearest, was that nobody deduced I needed immediate psychiatric care. On that horrendous day, there was no ambulance, no hospital, no 72-hour psych evaluation. There was just *prescribe her some more drugs* after a trip to the university clinic doctor. I highly doubt these doctors received copious amounts of training in this sort of situation.

Paradoxically, somehow, since I survived the death of a child, people just assumed I would survive the death of another child. Why not? The Marine Corps Sergeant Torrance suddenly popped into my head. He always told me, "Never *assume* anything—that only makes an *ass* out of *u* and *me*."

Did these people think the death of one child trained people for the death of another child? This was not even rational thinking for them, was it?

I certainly had no rational thoughts in my head. Sadly, people could not hear the deliberations that raced around in my brain. If they could have, there would have been different decisions made instantly. They would have put me in a straitjacket and taken me at once to a psychiatric hospital.

Psychiatric intervention, along with a long-term hospitalization, should have been the first thing done, at the very least. I could not find any solid ground to stand on or even a place to function from. There were no rules in my head for the death of my two sons; not even rules learned as a child from the Marine Corps Sergeant seemed to help me.

Where tragedy after tragedy has happened, never believe a person if they say they are *all right*. My mental state of mind went downhill quicker than an avalanche. Kim decided I needed another trip to the clinic at the university. Even though school was still out, Kim took me down the mountain to the clinic for another appointment.

The doctor rapidly realized I was not dealing with Bud's death. Once again, after no mention of psychiatric evaluations, the only answer they had was to increase the dosage of my anti-depressant and anxiety medication. I truly have no memory of the visit. Either I fooled the medical staff, or they were not comfortable with having me committed to a hospital for evaluation.

With the hell going on in my head, thankfully I did not decide to self-medicate with alcohol, but it was a constant struggle between not caring and not wanting to disappoint Bud and Jeffrey by drinking again. I easily could have drowned myself in alcohol. Miraculously, somehow The Promises I made to my children—this time to stay sober—proved more powerful than my desire to go numb with alcohol and to just lay down and die.

Police Update

Surprisingly, we got a brief update from the California Highway Patrol. It really was not good news, so it did not help my state of mind. Kim took the call and put it on speaker as I would not talk to them.

The officer said, "We wanted to let you know the car that hit Bud was found abandoned in Chumash, California, parked in a subdivision. The people living there saw the story on *Secret Witness/Crime Stopper* and realized the abandoned car sitting in front of their house was the car we were looking for, so they called us. We fingerprinted the entire vehicle. We now have everybody's fingerprints on file. We can only hope one of them applies for a job that requires fingerprints or commits another crime and gets caught."

Kim thanked them and hung up.

"They can only hope. What a freaking joke," I said.

Was the California Highway Patrol still trying to find the people that killed Bud? In my mind, I had resigned myself to the fact that they would not find these people. If only they had left one officer to watch the house, the perpetrators would be in prison.

These undocumented immigrants had outsmarted the California Highway Patrol; the police missed their chance to catch the people who senselessly killed Bud. For all the police knew, these people had left the country. Having lived through the murder of Jeffrey, I knew there was no such thing as justice in the United States *unjust* justice system. There certainly was no closure when it came to the violent, senseless loss of a child.

Psychiatric Discussion

After receiving more bad news from the police, Kim realized I needed to talk to somebody. She believed I was on the edge of doing something drastic. Kim suggested I talk to Dr. Carlson again and let her know what happened to Bud. Kim called her and explained the disastrous events of my life. This saddened and shocked Dr. Carlson. She was patient and spent time talking with me.

"Kathy, I am so sorry this happened again in your life. I cannot believe you lost Bud. He was such a nice young man."

"Yes, Dr. Carlson, it is so unfair."

"Kathy, are you suicidal?"

Of course, I lied. "No."

"You need to remember all the things we talked about. I know you have the tools we worked on after Jeffrey's death. They are the same tools that will get you through this crisis again. Bud and Jeffrey are together. They must have needed each other. I am so sorry you feel so alone. They are always with you."

It was helpful to have confirmation from her as a professional.

"Kathy, you really do need to write a book that tells the story of you and your children's lives. Other people need to hear what happened and how you survived."

"I don't know if I am going to make it this time, Dr. Carlson," I said.

"Kathy, you need to continue to honor your children. Do the things they can no longer do. Live your life and make them proud. Finish your education, like you promised them. Please, take the time to write the book when you can. Believe me, people need to know they can survive and even thrive after the loss of their child or any loss for that matter."

"I'll get to it someday. I promise you I will write it one day," I said.

We said our goodbyes. It was good to talk with her again. Nothing would bring Bud and Jeffrey back to me, but I knew I had to hold on, go to school, and survive. I was not sure at all that I could make it through the nightmare again.

BUD

Thanks to Kim!

Was any of this even remotely fair to Kim? She met me, and within three years, I lost both my children and my granddad. Kim lost her mother. The devastation was unbelievable. That was four deaths in three years. That *Soul Train* could stay far away from me. I never wanted to hear that train roar into town again.

Kim helped me get through so much wreckage in my life. It certainly had not been an easy road for her. Why she stayed baffled me. Why would anyone want to deal with me and my mental state of mind at that time? Kim deserved happiness, not death and doom.

The loss of my children was not Kim's responsibility. She deserved to be happy and have somebody who could be a part of her life, not my messed-up world. I had barely survived Jeffrey's murder. Bud's death was a whole new impasse to deal with. There was no guarantee that I would ever come back from this added loss.

Making Plans

As my brain swirled in a vortex of pain and agony, an idea developed in my brain: I would end my misery. I started researching the perfect spot on the mountain. I needed a big cliff with no barrier to stop my car as I drove off the top of the mountain at a high rate of speed. This thought began to consume me. The tricky part was I needed a 99 percent guaranteed probability that I would not survive the crash at the bottom of the mountain.

As I drove around the backroads of the mountain, I checked out various areas. If I found an area that looked promising, I parked, got out of the car, and walked to the edge to see how far down my vehicle would go before it hit something. Most importantly, whatever happened, this needed to look like an accident. I want to make sure that Kim would have my insurance to take care of her for years to come. It was extremely important to find just the right spot to make this accident happen.

Ironically, simultaneously, while I planned my demise, I kept Dr.

Carlson's words in my head, which was a direct contradiction to the above thoughts—my mind never stopped racing. Day or night, my mind tried to work out all the questions that swirled in my brain. *Why has all this horror and death happened? How shall I plan my demise?* At the same time, another part of my brain tried to figure out how to survive and honor both Bud and Jeffrey.

I tried to focus on The Promise I made to Bud and Jeffrey. It was amazing that my brain did not just implode. Amazingly, school would start again on September 26. I thought we had broken the trend of the 26th with Bud's death occurring on August 28, but it continued.

The Next Day Came. The days just kept coming.

35

LETTERS AND AWARDS

Two unexpected letters arrived in the mail. Although they were both in honor of Bud, they felt like another slap in the face. They were a reminder of a son I would no longer see or talk with. I know they were not meant to be hurtful, but I was not prepared to deal with anything more at this time. To me, a form letter was not enough thanks for the life of my son.

The first letter to arrive was from the Office of the President of the United States. When a service member dies on active-duty, their immediate family member receives a letter from the president thanking them for their loved one's life and service to our country. This letter, signed by President George W. Bush and framed, hangs on the wall with Bud's Navy awards and photos.

The second letter I received was from the Department of Defense. Inside the envelope with the above letter was a Gold Star lapel pin. This letter did not explain to me that the death of Bud automatically made me a Gold Star Mother, nor did it explain what being a Gold Star Mother even meant for me.

Kim bought a beautiful wooden memorial box for Bud's flag, medals, and other military paraphernalia. She stuck that Gold Star in Bud's case; we forgot about it for thirteen years. It was not until the fall of 2016 when, by mere chance, I met a Gold Star Father that I remembered that pin. One day, I walked into his restaurant which he decorated with all sorts of military paraphernalia and photos. After asking him why he decorated his restaurant that way, along with a conversation about our sons, he realized that I was a Gold Star Mother.

Before I left, he told me his wife was the president of the local Florida chapter of Gold Star Mothers, Inc., and he handed me a slip of paper with her phone number on it.

He said, "Give her a call."

I never called her, as I was still not talking on the phone, and certainly not to strangers.

Six months later, forgetting about the restaurant décor, I once again opened the door of the same restaurant. Before I could turn to run out, the owner called my name.

"Hey Kathy, have you called my wife?" He asked.

"No," I replied.

He took my order and nonchalantly handed me his already ringing cell phone call to his wife. Against every fiber in my body, we talked and agreed to meet for lunch the following day. She told me about the American Gold Star Mothers, Inc. organization, a national non-profit 501(c)3 corporation, with headquarters in Washington D.C. (To learn more about Gold Star Mothers, see the information at the end of this book.)

Shockingly to me, I learned that someone from American Gold Star Mothers, Inc. should have contacted me after Bud's death; they never did. The invitation—by mail or email—for me to become a member of their group never arrived. This group of women who shared the pain and sorrow of a child lost during military service stayed an enigma to me.

I continued to struggle for over thirteen years, unaware of an opportunity denied me, a chance to share my battles of grief with like-minded mothers. Would this have changed my healing process? I believe it would have been instrumental to know there were other mothers just like me with whom I could have shared my heartbreak and sorrow.

Sailor of the Quarter Award

Bud also received an award from the Navy just before he died. Commander Sawyer presented Bud with the VR-55 Navy Unit plaque for *Sailor of the Quarter* in dedication of his arduous work during his first tour of Operation Iraqi Freedom.

> According to the Navy, in April 2003, during Operation Iraqi Freedom, Bud's unit deployed three planes and thirteen crews. They flew 122 sorties, moved 1040 passengers, and 2.68 million pounds of cargo in theater.
>
> ~ *U.S. Navy News*

Bud joined the U.S. Navy on his seventh birthday, and his job in the Navy was as an aviation mechanic hydraulics (AMH), working on both the F-18 fighter and C-130 cargo plane. He served four years active-duty and was on his third year in the Reserves. Bud was going one direction and only one direction in his life; that direction was *up*, in both the Navy and his education.

I often wonder: *If the Navy had not reactivated Bud to full-time active-duty from the reserves for Operation Iraqi Freedom, would he still be alive? Could I have stopped Bud from continuing to serve in the U.S. Navy?*

Bud was my only surviving child. After the loss of Jeffrey, he was the sole male heir to his family name. Nobody answered my question. The brain game known as *would have, could have, or should have* remained a constant friend of mine. These questions made for sleepless nights.

Nightmares went on and on as my brain tried to find the

answers to these questions. To what avail? None. Nothing would ever bring Bud or Jeffrey back. The wars in my mind went on and on, just like the wars in the world. Children continued to die every day for a war that made no sense to me.

What was the exact number of children and civilians who had to die for the rich to make enough money from oil? *If we kill them, will they not kill us?*

The Next Day Came. The freaking days continued to come.

36

THE PROMISE TO KEEP

Every day continued to be a struggle for me. I could not find solid ground to stand on, nor a reason to continue with my life. Mercifully, school started again on September 26, 2003. *Is the repeating date of the 26*th *a message sent from Jeffrey in heaven? Does the 26*th *keep appearing to remind me that Jeffrey watches over me, loves me, and misses me?* Yes, I still believe this and hold it dear in my heart.

How will I be able to attend school less than a month after the loss of my second child in two years? Why do I put so much pressure on myself?

The only reason to continue college was the fulfillment of The Promise to Bud and Jeffrey. Those two little boys in heaven believed in me. At least going to college gave me something other than death and dying to focus on. School work gave me something to do, and going to classes occupied a tiny space in my brain that gave me brief moments of peace from the hell I swirled in.

Kim was skeptical about me driving up and down the mountain alone. Against her advice, I drove myself down the mountain to the campus, picked up my books, and checked in at the clinic.

More Medical Issues

As soon as I arrived at the clinic, the medical staff checked my blood sugar—it was 226. They were not happy with that number at all, as it was nowhere near the readings I usually had, around 120. Anything over 80 was not good. Diagnosed with diabetes after the loss of Jeffrey, the medical staff gave me a shot of insulin and increased my diabetic medicine to bring the levels back down.

My blood pressure was also high, again, so they decided to put me on a blood pressure medication. Stress continued to take a huge toll on my body. The doctor added more prescriptions for high blood pressure and high cholesterol after Bud's death.

The depression was relentless. Thankfully, I did not drink alcohol any longer, but food became a replacement for alcohol. I have learned this behavior is quite common for people with addictive personalities; they just replace one addiction with another. I continued to eat whole pans of brownies and all sorts of things I should not eat, trying to ease the pain.

Unfortunately, continuing to eat my way through my depression and anger caused further weight gain. Complicated grief, along with PTSD, took me into the depths of depression. Lacking focus, I found it hard to explain any rational reason I was still on Earth. No matter which way I looked at my life, I missed my sons. That was the fact, plain and simple.

School Problems

Back at school, my classes were Principals of Macroeconomics and Business Law. I had no idea how I would be able to go to college without Bud, but I had no choice. Bud was in heaven, he was dead, and I lived in hell.

Things went wrong from the moment I walked into school. I attended my first Business Law class. The auditorium was full of over two hundred students. The professor started to speak. I could not understand a single word he said. It was like he was speaking a foreign language.

BUD

All I could think about was Bud and how he should be there with me. In my head, I saw Bud in a cold dark casket in the ground. Bud had so many dreams of achieving a college degree. He loved learning. Life sucked; it was so unfair. I tried to focus, but I could not concentrate at all. I cried. There was no reason to sit in this class as it wasted my time. I excused myself and went home. I shared what happened at school with Kim.

"Kathy, just go back to school for your next class and keep trying. It will get better. You just lost Bud. It will take time. You know this is what Bud wanted you to do. The boys love you and are watching over you."

Later in the week, I went to my other class, Principles of Macroeconomics. The professor talked, but nothing he said went into my brain. *How am I ever going to make it through this?* I cried.

Marine Corps Sergeant Advice

Not wanting to go home again, I took a deep breath and tried to focus. Suddenly, a voice popped into my head. It was Torrance, the Marine Corps Sergeant. *Buck up kid. You have those huge crosses to bear.*

"Whatever a**hole, I'll do this for Bud and Jeffrey, not a crazy belief you have about me," I said aloud to nobody.

I made The Promise to those two little boys, and I would keep it. I sat up straight in my chair and took a couple of deep breaths. I listened closely to the professor, and it worked. I focused and took notes. Somehow, I made it through without crying. Surprisingly, it turned out that I was exceptionally good at economics.

Plans Changed

The next week arrived, and I went back to Business Law again. I took my seat, the professor spoke, and I bawled. This was not going to work at all. I could not take it. I left the classroom. I went down to the administration building and withdrew from the class. This was the first and only class I ever withdrew from in college.

The most significant thing to me was that I did not quit college, just Business Law. I continued going to Macroeconomics. It would have been easier to just quit and admit defeat. Beaten down farther than most people could have withstood, it took sheer determination for me to keep lasered onto The Promise, the one thing that was extremely important to me. Everything was just too raw; I could not tell up from down. My only hope was to stay focused on one class. I had to make it.

My health continued to deteriorate. The migraines were especially horrible as I experienced nausea and blurred vision. A migraine always presented on one side of my head and focused on that side's eye. There were times I sat through class, covering the eye with my hand, on the side of my head that the migraine affected. It made it hard to take notes, but I did not give into the pain.

Thankfully, I made it through the quarter and received an A in Microeconomics.

Funny Economics Story

The instructor, after learning the story of the devastating demise of my two sons, thought I was a miracle. He could not believe I was still alive and able to attend college. I told him I had promised my sons, so I had to follow through. He thought this made me special. I did not agree but listened to his thoughts.

"You should join the Peace Corp and help other people. You have unbelievable strength, and you must have special powers or something. How do you continue to walk upright on Earth, come to school, and do your homework after all that happened?"

"The only thing that keeps me going is The Promise to my sons. I live every day to make them proud and be the person they believe me to be," I replied.

I would get a college degree, no matter what it took, but I knew it would not be in business. There were just too many memories and thoughts of Bud. I knew I would not make it if I stayed a business major.

Lost and Found

Wondering around the campus, I ended up in the Criminal Justice building. I walked in the door of the Dean's office, and with no surprise to myself, I cried. It surprised the hell out of the people in the office, however. I wondered how often a student had walked in the door and started to bawl uncontrollably. Well, depending on grades, it might have happened more often than I thought.

I explained my situation to the lady at the desk, who introduced herself as Maria the office administrator. Maria did not panic. She called one of their professors to come out and speak with me. The professor introduced herself as Donna Primrose, PhD. After she calmed me down, I told her my story.

"I am so deeply sorry for the loss of your two sons. Your story is so profound. The loss of your children must be unbearable. We would be honored to welcome you into the Criminal Justice Department," Dr. Primrose said.

"Thank you; that is sweet of you. I was a law enforcement officer with a large department back in the eighties. I have no interest in becoming a law enforcement officer again, but I would be interested in learning why people commit crimes."

"This program is not about the minutia of a police officer's daily duty. The course goes into the psychology and philosophy of why people commit crimes. I was previously a police officer with a large department, too; we will have to compare stories. I still work as an officer at a local college. I think you will like the classes."

"Thank you. This is really important to me."

"I'm looking forward to getting to know you and more about your story. Go down to administration and change your major to Criminal Justice. Sign up for a couple classes, and we'll see you January 3, 2004."

We formed an instant bond and are still friends today. I liked Dr. Primrose. I hoped Criminal Justice might be where I would find out why people killed my children.

The Next Day Came. They just kept on coming.

37

A DAY TO HONOR BUD

On November 6, 2003, Mission College invited me to speak at their Veterans Day ceremony. They held a special ceremony to honor students who served in Operation Iraqi Freedom. Never having spoken at an event such as this—along with not speaking to people in general—speaking to a huge crowd of dignitaries and students was a huge step for me. Somehow, I knew in my heart that this moment was not about me; it was about honoring Bud and his service to our country. I decided to do this for Bud.

There was a wonderful memorial for Bud in the program.

In Memory of Bud
AMH2, United States Navy
July 17, 19** – August 28, 2003

Academic Achievements
Full-Time Dean's Honor Award: 4.0 GPA

After serving his country honorably, Bud enrolled at POMC and worked in the Financial Aid office and the Veteran Affairs program.

He was instrumental in organizing the Veterans Day ceremony following 9/11, and he served as guest student speaker.

Bud achieved considerable academic success until the U.S. Navy recalled him to active military duty to serve in Operation Iraqi Freedom.

He remained active in the Naval Reserve until his premature death on August 28, 2003, killed in a homicide on his way to NAS Point Mugu in Oxnard, California. His death shocked everyone who knew and loved him.

Bud's first love was his Harley-Davidson motorcycle, and he was known to say: "If I die while riding my bike, I died happy."

> *You're in the arms of the angel*
> *May you find some comfort here...*
> ~ Sarah McLachlan, "Angel"

Bud's Navy friends from NAS Point Mugu came to the ceremony to honor Bud. There were dignitaries who spoke. I was not sure how they knew about Bud, but they all had wonderful, kind things to say about him. A congressperson presented a flag to me in honor of Bud's service that had flown over the state capital. I know Bud would be proud and happy that all these people came out to honor him.

An American flag has flown over our home since Bud's death, no matter where that home might be. This is one more way I show my respect and admiration for Bud and his service to this country.

I spoke to over one hundred attendees in honor of Bud. I told of his love of his family, education, his commitment to help others, especially veterans, to achieve their educational goals, his love of the U.S. Navy, and his passion for Harley-Davidson motorcycles.

Kathy Speaking at Bud's Veterans Memorial

Reflecting on this event, I realize my speaking career began this day. It was the first time I shared Bud's story. I also included Jeffrey's story and mine, as our lives intertwined. The story of *The Three Musketeers* came back to life through my words. Was this also the beginning of *The Next Day Came Trilogy*, revealing our stories?

The possibility of a new purpose in my life peeked in every so often.

After the ceremony, we spent time with Rayna and other friends of Bud. It was wonderful to see them again and to hear more stories about Bud. I loved to hear all the stories, even if I had heard them before. He truly made an impression on VR-55, his unit at NAS Point Mugu.

This day had been a good day spent honoring Bud, but it ended and, as always, The Next Day Came.

38

LIFE GOES ON, OR DOES IT?

How could it be possible that our lives changed so drastically in such a brief time? In two years, the annihilation of my life as I had known it was complete. There would never be another holiday or birthday shared with Bud or Jeffrey. Everything in my life had gone so wrong. All my hopes and dreams—everything I worked so hard for—ripped from my heart, left me empty.

Kim and I decided not to decorate or exchange gifts during the holidays, ever again. Well, I decided the *ever again* part, and Kim went along with it. This was not fair to Kim, I knew, but it was all I could manage. We learned the hard way that life was too short to wait for holidays.

If we needed or wanted anything, we bought it. Material things were not important any longer. We understood that we did not know whether we would live long enough to wait for a holiday to get that gift we wanted or needed. We understood the importance of every day.

We looked at holidays so differently. *Celebrate what?* We discussed the situation and created a new plan. We decided to arrange a dinner each holiday for people we or our friends knew who would sit home alone during a holiday. We supplied everything, and we always set a place for Bud and Jeffrey at the table.

We had senior friends whose children did not live nearby. We shared holidays with friends of friends, single moms with children, spouses, and even children of friends whose partners or spouses were traveling. We even had a girl who went to the high school in the small town where I grew up, who was newly divorced. She brought her dog.

Since we could not be with Bud and Jeffrey for the holidays, we found a way that gave back and shared the holidays in their honor. Kim, raised Catholic, said a prayer before dinner, and I shared a story about Bud and Jeffrey, so people understood the purpose of the dinner. This tradition continues even today. We have fed as few as five people and as many as twenty-four. It makes the day a little less dreadful.

Pain and Personal Possessions

Kim decided after the Christmas holiday that we should tackle Bud's stuff, down the spiral staircase. The belongings had sat there since September 1 when the Navy delivered them, along with Bud's truck.

Bud's truck still sat in the driveway, as we waited for the bank to catch up with us. Kim was sure there was paperwork in Bud's belongings that would tell us who to contact. I had no desire to go through Bud's belongings, but Kim felt they at least needed organization, so around and down the spiral stairs we went.

Mostly, I cried. Kim gathered and sorted all of Bud's clothing, as in her mind, they needed washing. We found all varieties of paperwork. Dozens of papers looked important, and others were memories for me to cherish. Kim put them all into a pile for her to go through.

Surprisingly, we found a semiautomatic 9 mm pistol. I forgot Bud bought it when he lived out on the East Coast. I wondered if the Navy even knew they loaded it in his truck and delivered it to my house. I was sure that Don knew, and he had placed the pistol in the truck.

Let me just say it was one horrible, tough job, going through my dead son's personal belongings. At twenty-four, he had way more stuff than Jeffrey had at eighteen. Besides, I remembered, somebody had stolen all Jeffrey's stuff.

All of Bud's possessions would need to be gone through again. Kim packed things in plastic tubs. She felt more comfortable with Bud's personal belongings organized and cleaned. The whole disaster broke my heart and made me cry. Any other decisions would need to wait until later.

Celebrating Bud's Navy

A fun thing happened on January 7, 2004. The Navy again invited Kim and me to take part in the 2nd Annual Bud Golf Classic. It was wonderful to see the people at NAS Point Mugu, although it made me cry to see the boys in their uniforms.

They held the tournament once again at the base golf course. Commander Sawyer played in the group with Kim and me. Having grown up playing golf, Kim had played at pristine golf courses. She was a little shocked by the condition of the course at the naval base.

Kim entertained everyone, as usual, and asked dozens of questions. Halfway through the round, she had a question for the commander that made everybody laugh. "Do you use the golf course for bombing practice?" Kim asked.

The commander dropped his golf club and almost fell over laughing. "No, although it does appear that way, doesn't it?"

The golf course did have huge holes in the fairways, but it was fun to play golf with the commander again. I cherished every moment with these people and their memories of Bud.

After golf, Kim and I had lunch with Rayna and the women, then shopped at the base exchange. I bought two pairs of Navy sweatpants, which I wore until they fell apart, along with a couple of shirts. We also found these little Navy sailor heads with sunglasses to hang from the rear-view mirror in our cars. Kim and I both have one in our cars even today. These things remind us both of Bud every time we see them.

Navy Guy Head

Change of Majors

Thankfully, school started again on January 9, 2004. I signed up for two classes. I would pursue a Bachelor of Arts degree in Criminal Justice. My classes were Alcohol, Drugs, and the Criminal Justice System, along with Criminal Procedures.

I continued to struggle with depression and stress-related illnesses that took a toll on my overall health. School was the only thing I could focus on, and that was overwhelming most days. At least I had something else to think about besides death and my lousy life.

I had The Promise to fulfill. I must admit it was easier to focus on criminal justice. Not having the constant reminder of the absence of Bud like I did in the business courses gave my brain a chance to concentrate on what the professor spoke about. The biggest change I noticed in this new major was I did not sit in class and cry. Another benefit of these large classes was that I did not have to talk.

It took years for people to notice that not talking had become a habit for me. I only spoke when asked a direct question. In fact, not talking on the phone and even in person went on for ten years.

Criminal justice courses held my interest, which made college a little easier for me. I received an A in both courses. These classes ended on March 15, 2004.

Angelversary and Alcohol

The eleventh anniversary date of my sobriety rolled around on March 21, 2004. That I remained sober was a miracle. Five days later, the reality of Jeffrey's death hit me like a hurricane on March 26.

My sobriety date, March 21, and Jeffrey's Angelversary, March 26, would forever merge. A week of what should have been a celebration turned into a nightmare from hell. With the added loss of Bud, it would have been so easy to take that first drink. I thought: *What difference will it make? Why bother to stay sober?* I only had The Promise to my sons. Hundreds of my days were certainly harder than others.

These constant reminders came around again the same time every year. I had to get through their birthday, their death day or Angelversary, and holidays all year long. It was an endless struggle.

I never knew what would be the straw that would break the camel's back, resulting in my breaking down and bawling. It could be a song on the radio or something in a movie. One thing that always made me cry were scenes of graduations or weddings. For the longest time, Navy commercials made me cry.

Pet Peeves

One thing that broke my heart and made me extremely angry—and truly caused me to bite my tongue—was when people complained about their children or grandchildren. They had no idea how extremely blessed they were, and they took those blessings for granted. What I would not give for the gift of having my sons or even grandchildren!

An example of this happened one day when I went to my usual beauty salon to get my hair done. I sat in the chair, and as the stylist worked on my hair, I listened to two women talking to each other. It was a non-stop conversation, or bitch-fest, for over an hour. One of the customers complained about her daughter. She went on and on about how her daughter did not do this or did not do that. Finally, the other lady started talking. I thought the conversation would get better, but it did not.

The other lady complained about how she had to babysit her grandchildren and how horrible they were. She went on and on about how it interfered with her life and how ungrateful her family was. It got downright nasty at times. I bit my tongue until the stylist completed my hairstyle. I paid my bill, and then I walked over to the two women.

"Excuse me," I said. "I sat here listening to the two of you bitch and complain about your children and then your grandchildren for the last hour. You should realize how lucky and blessed you are to have children and grandchildren. People killed my two sons in separate homicides. They were my only children. I will never have grandchildren. I am here alone. So, next time you feel like bitching about your family, you might stop and think about what it would be like to lose them all and be thankful for the gifts you have been given."

They stared at me as if I had two heads. Then they lowered their heads. Neither one of them said a word. I hoped it hit home, and they would decide to change their lives. The two hair stylists smiled broadly. They must have been tired of listening to the two women complain, too.

"Thank you," they mouthed, in unison.

I walked out of the salon, got in my car, and cried. *How will I ever survive in this world without Bud and Jeffrey?* I was sick and tired of being in pain and miserable. This event pushed me over the edge.

How could people be so ungrateful and selfish? They had no idea how dreadful things could get, nor how quickly.

What I would not have given to have my sons and be able to talk to them again. My thoughts rolled. *Why am I still here? Why can I not just die and see Bud and Jeffrey?*

The Next Day Came. The freaking damn thing would not stop coming.

39

PERMANENT SOLUTIONS

With the arrival of another March 26 in 2004, a deep and dark depression set in. I truthfully could not take it anymore. Horrible thoughts swam in my mind like a shark circling its prey. The philosophies in my head filled with hopelessness. I laid there and thought of ways to leave this Earth. I rarely slept, and when I did, I had unbelievably graphic, atrocious nightmares of death and suffering. With nobody to talk to—at least I felt that way—I existed in silence.

Looking back, I should have been in psychiatric treatment from the day I learned of Bud's death. Even though I was a patient of the psychiatrist Dr. Carlson after the loss of Jeffrey, the death of a second child, my only other child, pushed me beyond what I knew how to deal with. In my mind, I thought that Dr. Carlson had taught me enough, but I was wrong. Talking to a professional who could have helped to clarify my thoughts, feelings, and beliefs with this second loss in two years, certainly would have been beneficial for my health. As the saying goes, *Hindsight is 20/20.*

Dinner and Confessions

Kim and I went out to dinner for Jeffrey's Angelversary as we had for the last three years. This time, I picked a fancy steakhouse. The cost did not matter; nothing mattered. The dinner was to honor Jeffrey.

I tried to teach Bud and Jeffrey the enjoyment of a delicious meal at a classy restaurant. Wyatt would dispute my ability to do this, as I was a fussy eater; I still am today. Having grown up with everything cooked with a can of Campbell soup as an ingredient, it took me years to learn to appreciate a delicious meal. I have remained a fussy eater, but I enjoy different cuisines that I never ate growing up.

Before Kim and I left for the restaurant, I had already decided it was time to share the truth with Kim about my thoughts and plans. I waited until the waiter served our meals, and then I brought up the subject. "I wanted you to know that this has not been an easy decision, but I have thought long and hard about this, Kim. I do not want to be here on Earth any longer. I researched and planned my demise. My only concern is that I do it correctly, so I do not end up in worse shape than I am right now. I've figured out a way to make it look like an accident, and I'm 99 percent sure I will not survive."

Kim said nothing, she just stared at me. I cannot say she was surprised by what I said, but she had not expected me to say the words aloud.

"I am so very tired, Kim. I struggle to make it through every day. I try to do the right things, but there just doesn't seem to be any point to doing those things anymore."

"Kathy, you have to give this more time. It has only been six months since Bud died. You have not even gotten through the loss of Jeffrey. Now with Bud, it could take years for you to feel better."

"Kim, you've always been there for me and helped me, but there is nothing I want to do. There is no place I want to go. Truthfully, I have done all the things I wanted or needed to do. I just want to go home and be with my sons."

"I know you do, Kathy, but I don't think that is what God has planned for you."

"Kim, you know I don't believe in organized religion."

"Well, my God has huge plans for you, Kathy. He would not have put you through all this pain and suffering if he did not have a huge purpose for you to fulfill."

"Your God and his plans have not worked out too well so far, have they?"

"No. It hasn't made sense to me either. I don't necessarily understand it, but I know that you are not supposed to leave here yet."

"I am just so tired. I appreciate all that you have done for me. You blessed my life with kindness and understanding. I don't know why you stayed with me."

"Because I love you."

"Look Kim, you're a wonderful person and you deserve so much more. You deserve to be happy, to laugh, and have fun. You need somebody to share and be a part of your life with you. I have nothing to give. They've taken everything I have."

Kim took a deep breath. She thought long and hard before she spoke. "Kathy, listen, you've had horrible, unbelievable events in your life. You have been through things that no parent should ever have to endure, much less twice in two years. I know you have a broken heart. I see the sadness in your eyes and face. I am not sure how you get up every day, but you take that leap of faith and get through the day. I sure am glad that you do."

"That is what I am trying to tell you, Kim. I am done."

"Kathy, this has aged you so much. I worry about you every day. You need to hang on and take time to figure out why you are still here. Your purpose will come. Just survive for right now. There is an answer. I know you are supposed to be here."

"I don't know. I hurt so much. I feel so alone. There is nothing but emptiness in my heart and soul. You are the only one who cares whether I am here or not."

"People love you. Why do you say that?"

"Really Kim, everybody went home and got on with their lives. I have no life. I never hear from my family. They never call or ask about how I am doing. They do not even bother to call on the day Jeffrey died, or on Bud and Jeffrey's birthday. I bet you they will not call on the day Bud died either. It's as if my children never existed to them."

"Kathy, I know that breaks your heart even more. If I could fix it, I would. I don't know why people do or don't do things. They're afraid, I guess."

"People are idiots. They do not realize the extent they hurt other people by not talking about their losses. I certainly have not forgotten that Bud and Jeffrey lived or died. They would not be reminding me about it. I live with their loss every freaking day!"

"They don't know what to say, so they say nothing. That does not make it right, but you have an opportunity to show people they need to do better, not hurt the people they love even more."

"Kim, my life is already devastated. Why do I always have to be the one to fix my freaking family and other people? Who is ever going to fix me? How would I even show them?"

"Maybe that is your purpose to still be here. You could write that book Dr. Carlson is always asking you about. At least think about it," Kim said.

"Okay, I'll think about it. You know I promised Bud and Jeffrey that I would get a college degree. I do my best, as you know. I get perfect grades in all my classes. I do everything I can every day to make them proud of me. I cannot write a book about all this yet, but someday I will."

"I know they are proud of you, Kathy; I'm proud of you. I know they look down from heaven and smile at how you excel at school."

"I guess. I am just so tired of the pain. I don't know how to put it into words that you would understand. There is nothing but dark emptiness inside me."

"Will you do me this one favor? Will you promise me that you will hang around for now? I will do everything I can to support you and help you through this. Please don't leave me. I need you. Okay, promise?"

"I will always be honest and tell you where I'm at mentally. I will not do anything without you knowing first."

Kim looked at me and smiled. "That's all I can ask of you. I need you to hang around. What would I do without you?"

"Well, I'm sure you could find somebody who wasn't sad and depressed all the time."

We finished Jeffrey's Angelversary dinner. It was delicious; the steaks were rare and succulent, just the way Jeffrey loved them. He would have enjoyed his Angelversary celebration, even if he had concerns about his mother's thinking.

My integrity would not allow me to kill myself without first explaining why to Kim. Telling her turned out differently than I thought it would. I still had no idea what my life was supposed to be about. Then again, nothing in my life ever turned out the way I thought it would.

Kim and I headed home. I made a promise to Kim, and I kept my promises. "Kim, I won't do anything to expedite my exit from this Earth, but I will not do anything to prevent it either."

"Fair enough. I'll take that," Kim said.

The Next Day Came, just like it always did. I could not get them to stop.

40

THE PROMISE

School started again on March 29. I noticed a couple young girls kept appearing in the same classes I was in. We were on the same track or working towards the same degree, I guess. Their names were Julie and Erin. The three of us always sat in the same seats. Erin and I sat next to each other in the front row of the class, and Julie sat behind one of us.

The realization hit me that I sat in the front row of the class the entire time I attended college. If I had to sit anywhere else in the room, I was unable to focus and learn. Otherwise, I spent my time looking at the other students, wondering about their lives. I started asking myself: *Why are these young people still here when my sons were both dead?* Lost in those thoughts, the class would be over before I realized I had missed the entire event. Sitting in the front row forced me to focus on what the professor said and nothing else.

I learned that Julie was twenty-two and single. Erin was twenty-four and married, with no children. They were around the same ages Bud and Jeffrey would have been if they were still alive. The three of us became good friends and started to spending time together between classes and eating lunch together.

It was comforting for me to have these two young girls around.

We also started to study together. Eventually, I told them the story of my two sons and their deaths. They were shocked and deeply saddened. At least they understood why I rarely talked to anybody.

On a mission, I took copious notes, completed my assignments on time, and did my absolute best. My classes were Research Methods in Criminal Justice, Correctional Theory, along with Institutions, Women, and Crime.

The other students soon realized that I received an A in all my classes. Dozens of them got a C or D and told me these grades were *good enough*. I laughed and told them the story about my youth. "Growing up at my house, you earned an A in all your classes or things went bad quickly. At the least, my parents grounded me for weeks. At worse, my father's leather belt came scorching out of his pant belt loops like a flame thrower."

> *Kids today don't know the fear of hearing leather being rapidly pulled through seven belt loops.*
> —John Wayne

Mentorship

Julie and Erin encouraged me to start a study group. This was a struggle at first, but I decided I would help other students in honor of Bud and Jeffrey. This opportunity gave me a new purpose to focus on. As the students slowly became interested, I mentored them with their studies.

We decided to meet before each exam for a review. I shared my study habits and gave them a copy of my notes. Miraculously, the students who attended the study group started getting better grades. The smiles on their faces were worth my discomfort in talking to other people. These better grades made us all proud.

Giving back to others allowed me to honor my two sons, which opened my heart to heal. If I could not help Bud and Jeffrey anymore, then I would help these students. This study group continued throughout my time at the university.

I missed Bud and Jeffrey so very much. At times, during classes, it made me incredibly sad. I knew Bud wanted to be there so badly. I thought about how he was the reason I was in college to begin with. I focused on my studies and did the best I could. I dedicated myself to keep The Promise.

Summer School

Time at school made the day pass by. I received an A in my courses. Since I had nothing else to do, I signed up for summer school. My next classes were Theory of Crime and Delinquency, Statistics in Criminal Justice, along with Police and Police Systems.

Statistics quickly baffled and frustrated me. To me, statistics was like learning a foreign language. It made absolutely no sense to me at all, and I struggled. This was truly a new concept to me, as school had always come easy to me. Growing up, I was always excellent at mathematics and enjoyed them.

Looking back at my high school years, I realized I never had to work at school. I had a photographic memory. If I read the material, I could flip through the pages I had read in my mind and find the answers when it was time to take the test. This process made school easy, and I always received good grades.

As an adult, I knew to ask for and received extra help from the professor. Eventually I got the hang of statistics. It genuinely surprised both Kim and me how the process of number manipulation could easily make people believe whatever the manipulator wanted them to think. This gave me a whole new insight into politics and all the statistical numbers I heard on the news.

My other courses were interesting, and I enjoyed them. The study group continued to gain in popularity, even in summer school. Thankfully, school and homework kept me busy and filled my time. This distraction forced me to focus on something besides death.

Changes in Thought

One day I woke up, and it was like somebody flipped a light switch up. A light came on, and I realized that I was going to live, whether I wanted to live or not. I still contemplated dying. In fact, I prayed at night before I went to sleep that I would not wake up again. Unfortunately, the Next Day Came and I opened my eyes. On this morning, it became clear to me that I would not die of a broken heart.

I cannot remember what changed, if anything, but somewhere in the back of my mind, it became clear that I needed to figure out a plan for the rest of my life. My physical health, gone to hell with all the stress I was under, needed attention, along with my mental health.

My body weighed one hundred pounds more than the day Jeffrey died, thanks to weight gain that continued with the loss of Bud. A plan popped into my mind, and I registered for three physical fitness classes. My transcripts showed I took weight training, golf, and swimming. Unfortunately, by this time, my screwed-up metabolism refused to cooperate with my losing weight. Obviously, losing weight was not going to be as easy as it had been to gain.

Swimming Lessons

As a child, Melinda taught me how to swim. Well, at least she taught me how not to drown in the lake. Melinda took us to our summer home on Deer Creek in Wisconsin every summer. As the only adult, Melinda wanted to make sure we would not drown if we went into the water unsupervised. Sadly, she taught us to swim without putting our face in the water, unless we swam underwater.

There were dual motivations for me to take swimming classes. I wanted to learn how to swim like Bud and Jeffrey used to swim. Competitive swimming was what I wanted to learn. I wanted to learn the breathing technique that allowed for continuous graceful laps. I also wanted to learn how to do those fancy flips at the end of the pool that Bud and Jeffrey performed so gracefully. Bud, Jeffrey, and I had attended hundreds of swim practices and swim meets as they were growing up. It always fascinated me that they could swim for hours and never seemed to get tired. I wanted to learn something they both enjoyed and excelled at doing.

My friend Julie decided to join me in these endeavors. We showed up, even though neither one of us were ecstatic about being in our swimsuits. We went out to the pool and lined up with the other kids. Things did not go quite as I had planned.

During the first class, the instructor had all the students swim laps, so he could evaluate our swimming level. Julie and I both ended up in what the instructor called remedial swimming. I felt like a failure, but I would not let that defeat me. I would learn how to swim to make Bud and Jeffrey proud.

Julie and I quickly learned that the school heated the pool water, which was great. However, they could not heat the air in California. There were dozens of chilly morning swims. I knew I would never match the abilities Bud and Jeffrey had when they swam, but I was determined to learn the techniques.

Swimming was enjoyable, and it was good for my physical health—at least that was what I kept telling myself. I did eventually learn to do the breathing technique correctly. Even more shocking, I performed the flip at the end of the pool. Silly me, I thought it would help me lose weight—it did not. There was no doubt that Bud and Jeffrey enjoyed hundreds of laughs at my swimming abilities, but in heaven, they were proud of their mother.

Golf Lessons

Kim's father owned the local sporting goods store in the town where she grew up. Kim is left-handed, but because she was a girl, her father did not think she would take golf seriously. So, he would not buy her left-handed golf clubs. Kim learned to golf right-handed and proved her father wrong. She became an excellent golfer and loves to golf, still today.

With this story in mind, I decided to learn how to golf. Who knew that golf was a class available to take while attending college?

Surprisingly, golf became therapy for me. I went to the driving range, bought a $10 bucket of balls, and released an enormous amount of anger and frustration.

Golf was immensely enjoyable to learn. As it turned out, I was astonishingly good at smacking the little white balls at the driving range. As a child, I played thousands of hours of baseball and softball. If I could not find anyone to play baseball with, I tossed rocks up into the air and hit them with my bat for hours. For me, golf was like swinging that bat, which I realized had also been therapy for me. My childhood is another story, which I share in book three of this trilogy.

Another blessing of golf was that it became something Kim and I enjoyed together. Unfortunately, golf was not a cardio exercise, but it relaxed me. I cannot speak for the people who played golf with me. Let me say that I was not good at following the rules of golf. If I did not like where my shot landed, I just hit another ball. I always carried three golf balls in my pockets. Kim started calling my style of golf playing the *Florida Rules*.

It was clear to me that I was not going to win a million dollars at the game, so I played for fun and worked my frustrations out. On the golf course, Kim and I enjoyed just being out in Mother Nature. We saw hundreds of birds and animals.

Summer school ended. With extra help from the professor, I passed Statistics with an A. I hoped to never see a statistics course again. I also received an A in all my physical activity courses. I was another semester closer to The Promise of a college degree. Everything in my life revolved around The Promise.

The Next Day Came. They continued to come, over and over.

41

BIRTHDAYS WITHOUT BUD AND JEFFREY

Birthdays in heaven must be different than birthdays on Earth. I missed sharing these special days with Bud and Jeffrey. I used to tell people it was like having Christmas in December, and Christmas in July. They enjoyed great parties and received wonderful presents twice a year. My birthday in January was a dud. As a child, I constantly received presents at Christmas that my parents said counted for my birthday too. From a kid's perspective, this was the ultimate rip-off. I never wanted my sons to experience that feeling.

In heaven, I pictured birthdays as festive events with all the family members who had died in attendance. July 17th became a different birthday tradition of sorts for Kim and me. After Jeffrey's death on March 26, 2001, Kim and I did our best to help Bud through the common birthday he had shared with Jeffrey for eighteen years. Not surprisingly, Bud had no interest in his own birthday any longer. There was no time for me to mourn Jeffrey's birthday. Bud and I were the only ones who talked about Jeffrey.

After Bud's death, every July 17th was once again Bud's and Jeffrey's birthdays. The way Kim and I looked at it, if Bud and Jeffrey were still alive, we would have taken them to restaurants for their birthday dinner. Regrettably, we now go without them. My world would never be the same, but I found ways to compensate for the pain I lived shrouded in every single day. Bud's and Jeffrey's birthday and Angelversary days were all I had left.

Another year had gone by since that dreadful day on August 28, 2003, when I lost Bud. For Bud's Angelversary, Kim and I decided to go to Roy's Hawaiian Fusion Restaurant in Poison Oak. We had gone to this same restaurant with Bud and Marisol when we took Jeffrey's ashes to Hawaii in 2001. Kim made reservations for four, but as usual, there would only be two.

Since we were going all the way to Poison Oak, we decided to see a show while we were there. We had not been to a show since the fiasco with Cher's concert after Bud died, so Kim bought tickets for a Melissa Etheridge concert. Kim and I enjoyed a wonderful dinner, along with stories of Bud and the memories we cherished.

Realizations

Strangely, an odd thing happened when we arrived at the concert. Kim and I settled into our seats. The people around us were drinking, obnoxious, dancing, and already enjoying the event. After the show started, during the second song, Kim and I looked at each other and knew we just wanted to go home. We had absolutely no interest in the crowds or the music. Our lives had changed so dramatically, and things like this just were not important or pleasurable to us any longer.

Throughout the years, people have often asked me the same question. "Kathy, how did you ever manage to survive the loss of your two sons?"

They look at me oddly when I answer their question. "The Next Day Came; they just kept coming. Next thing I knew, years and years of next days had just passed me by."

People have told me that they do not know what to say to me; they do not want to remind me of the deaths of my two sons. The loss of my children never went away—that fact was in my brain from the moment I opened my eyes in the morning to the moment I fell asleep at night. I knew both my sons were dead, so nobody ever needed to worry about reminding me. Talking about them would have been a kindness to me, knowing they had not forgotten Bud and Jeffrey.

As I moved through these early years, thoughts galloped through my head: *What would my life be like if my sons were alive? Would I have grandchildren now?* I know in my heart that I would have been the best grandmother, but that was something I would never experience. *What would my career, or relationship with Kim be like?* Nothing was as it should have been. *Will I ever find that Different Place, Dr. Carlson?* At least I had Bud and Jeffrey forever in my heart.

I kept myself focused on school. That was my one and only goal. I had to get that college degree. I had The Promise to keep. I still did not communicate with people any more than I absolutely had to; I answered questions when asked. My main sources of people to talk to were Kim, my friend Bonnie, and Jerry across the driveway.

Soul Train **Roared Up the Mountain**

The first of September 2004, shockingly, doctors diagnosed Bonnie with stage four ovarian cancer. In her late forties, full of life, and appearing to be an extremely healthy person, the news outraged all of us on the mountain. The prognosis was not good at all.

Bonnie was a great friend to me from the time we moved to the mountain. She had listened to my stories about Jeffrey, and later Bud. As an attorney, she encouraged me to finish my college education. She knew how important it was for me to make my boys proud. She was a source of compassion and understanding for me.

Bonnie died quickly. At least she did not suffer. The *Soul Train* roared up the mountain. The gate crashed down over the crossing while the red lights flashed on the gates. *All Aboard.* The *Soul Train's* whistle blew, as Death #5 hit me while I stood on the track as the engine plowed me down.

This loss brought back deep, dark emotions from my memory. Horrible nightmares interrupted what little bit of sleep I was able to get. My thoughts avalanched deeper into the darkness of loneliness and isolation. My depression became steadily worse.

The Universe would not give me a chance to breathe, much less heal or find solid ground to stand on. Every time I felt just a little bit better... *Bam! They* took somebody else.

There was no way I could attend Bonnie's funeral. Death was just too raw in my life. Kim, along with our mountain friends, went to her service. With the recency and rawness of Bud's service, I thought I might never attend a funeral again.

Another Promise Made

Before Bonnie's death, I promised her I would donate my long hair in her name to the organization that made wigs for cancer patients who lost their hair to chemotherapy. This allowed me to honor Bonnie in my own way. I kept my promise to her and donated my hair three separate times in her name.

It had been a tough summer. Exhausted, depressed, barely functioning at times, I wondered if this *Soul Train* would ever stop running. I wanted it to derail and leave me the hell alone. Better yet, I wanted the engine to go into the round house and turn the train around. *Retire the damn thing!*

Kim and I lost five people in four years. There were no words to express how tired I was of losing people. If I never heard about the death of someone I loved again, it would have been too soon. My capacity for pain and loss had long depleted; there was no more room for the devastation and trouncing I had endured. I wanted my children back.

My life continued to be *one step forward, and three steps back.*

The Next Day Came. Days just kept on coming.

42

THE NEXT DAY CAME —THEY CONTINUED TO COME

Thankfully, school started again on September 23, 2004. It was good to be back in classes again. I did not talk to people, but I enjoyed learning new things. School gave my mind something else to focus on besides death, and homework kept me busy. Giving back and honoring my promises helped me move forward, just a little bit. These things opened my heart little by little.

It was good to see Julie and Erin. Once again, we were in the same classes. My motivation was strong as I moved closer to fulfilling The Promise with every class. My classes were Correctional Administration, Hate Crime Law and Policy, and Integrative Studies in Criminal Justice.

Julie and I continued with our physical activities of swimming, golf, and weightlifting. The three of us continued the study group, and I noticed that more students joined the group. Before mid-term tests and finals, we met at our house on the mountain. Kim, always feeding people, made spaghetti and bread for everyone.

This progress encouraged me to continue the mentoring group. Their study habits improved; could this improve their lives too? We became a cohesive group of students. If I could not help my sons, I would help someone else's children succeed. Once again, I received an A for all these courses.

Holidays Again

When the holidays rolled around, Kim and I continued our new traditions. We opened our home, so people did not sit at their home alone. We supplied all the food and drinks, but a half of dozen people asked to bring their own special dishes. Kim agreed they could bring something or nothing at all; it did not matter.

We had about twenty people for dinner. Friendship and conversations made the day a little better. Everyone had an enjoyable time, or so it seemed. I thought about all the holidays that The *Three Musketeers* shared together. It was strange to be alone in a crowded room; I do not have the words to describe that feeling.

My life had changed so much. I spent twenty-five years of my life focused on giving birth to and raising my two sons. Even as a single parent, I always managed to give them the best holidays ever. Unlike my childhood, I made sure they received the things they wanted.

Christmas Stories

I always shared the stories of Bud and Jeffrey and why we held these annual dinners. Occasionally, I threw in a childhood story or two about Bud and Jeffrey. People were always surprised when I shared that they always received the exact same things for Christmas and birthdays. They received the toys they wanted, one for each, along with the same clothing, in their sizes. I was sure it would curtail arguments.

Melinda had a story of her own. She pointed out that for twenty-five years, I always did for my children first and myself last. "Kathy, now that your children are gone, are you finally going to do things, just for yourself?"

"Probably not. I don't need things," I said.

What an odd question for the holidays. Even today, I remain a minimalist person. *Things* no longer matter to me.

Christmas Trip

Kim and I made it through the holidays and decided to take a trip to see Kim's stepdad Skip. He lived in Grama, Arizona, in a lovely home at the Woolsey Country Club. His home sat parallel to the right of the fifteenth green on the golf course.

It was so peaceful to be there and watch the golfers play their games. From the porch, I watched the golfer's tee-off on the par three hole to the left and hopefully land on the green. Occasionally, the ball came into the yard, and we had to duck to escape injury from a wayward golf ball.

Skip was a wonderful man. He became Kim's stepdad when she was in college, as her father died after she graduated high school. Skip was a retired schoolteacher and band director in Wisconsin. He loved it when Kim and I came to visit. I am not too sure he was thrilled when we brought the two standard poodles with us.

One night, when Kim and I walked the poodles Jake and Keleli on the golf course, we ran into a bunch of wild Javelina. They stared at us; we stared at them. Jake and Keleli thought they should defend our honor or save our lives. Outnumbered, Kim and I were not up to a confrontation in the dark. We took off and made a beeline safely back to the house.

Golfing Adventures

Skip was an excellent golfer and, thankfully, a patient instructor of the game. We played a round of golf with him every day while we visited. I was not a particularly good golfer yet, but Kim played well. I had wild shots or no shots at all. Skip was encouraging. If the ball went forward, Skip cheered me on.

"Get in the cart, Kathy, and hit the ball again," he said.

Skip was happy to have us there, and it was a bonus that we were willing to play golf with him every day. I am sure he had nightmares after playing golf with me, but he never complained. Skip encouraged me to continue to learn golf. I was not sure whether that was a compliment or not. Playing golf, for me, was a break from the reality and hell of my life. Unfortunately, Kim and I had to go back to California.

Back to School

The bachelor's degree program moved along quickly. It was hard to believe there were only two quarters left. During these last two quarters before graduation, the program required students to help professors as student teachers. I worked with Dr. Primrose and taught three or four classes for her. It was a bit bewildering to me, but I enjoyed the teaching process. For a while, I even considered becoming a teacher. Teaching forced me to speak to people, but I only spoke about school stuff.

People wondered why I did not talk. If they asked, I told them how it came about and that I had nothing to say in person or on the phone. People had no understanding of what went on in my head, and I did not want to explain it every time, so I said nothing. Honestly, they had nothing to say to me that would have changed my circumstances or brought my sons back.

On January 6, 2005, school started again. This quarter my classes were Law and the Courts, Comparative Criminal Justice Systems, and Gangs. Of course, Julie and I continued our physical training with swimming and golf. We had a scheduling conflict and could not take weight training.

Julie, Erin, and I continued the study group. The popularity of the group grew exponentially. Students realized it was easier to work together, and participation in the group supplied them copies of my notes. They began to find a sense of pride in their better grades.

If I knew then what I know now, and had been more entrepreneurial, I would have sold my notes. Looking back, I realized that my notes were priceless. I was a great note taker and extremely organized. Had I sold my notes, could I have paid for my college education? Regardless, my work in school made me proud, and I was sure Bud and Jeffrey were also thrilled for me. Helping these students reminded me of caring for Bud and Jeffrey; I wanted these students to succeed too.

Unexpected Honors

The Criminal Justice Department at California State University won a research project for a major city in California, researching domestic violence—what worked and did not work. The department head asked me to research and write a literature review for the project. Even though I did not receive academic credit for the project, it was a great learning experience for me and an honor that they asked me out of all the students in the Criminal Justice Program.

Medical Issues

The stress-related diseases that developed after my multiple losses continued to take their toll. I had not slept through the night in over four years. I woke up every few hours and could not get back to sleep for hours. According to my doctor, lack of sleep was deeply affecting my health. The doctor explained to me that a person's body needed to go into a deep REM sleep at night to heal, both mentally and physically. My interrupted patterns made it difficult for my body and mind to begin to heal. The doctor suggested natural remedies to help me sleep, but they did not work. I was unwilling to go on another prescription medication to make me sleep. The lack of a good night's sleep continues, still today.

Besides, sleep brought nightmares or extremely vivid dreams in which Bud and Jeffrey appeared at various ages and in different situations. A funny thing became clear to me; these events in my dreams were not always memories, at least not memories I recalled ever happening in our lives. I found this extremely interesting, and I wondered if *The Three Musketeers* existed in a different realm of time and space.

These dreams were very realistic, as if I had just talked to Bud or Jeffrey. I enjoyed them, but occasionally these dreams were violent, especially if they involved Jeffrey. A strange phenomenon in my life became about time. In the world, time moved forward. In my broken heart and mind, time stood still. I learned: *Time does not heal all wounds—it is how you use your time.* For me, I spent my time on my schoolwork to fulfill The Promise. The rest of life passed me by.

Focused on The Promise to Bud and Jeffrey, I kept moving forward. I barely functioned other than school. The seclusion of the mountain allowed me to detach from the world. Mostly, I spent time with my standard poodles as they gave unconditional love and loved my stories.

I received letter grade A in all my classes. Unbelievably, the quarter ended on March 26, 2005, the fourth Angelversary of Jeffrey's death. I thought: *How can four years have passed without my Jeffrey? Where did the time go?*

I continued to lose track of days, weeks, months, and years. I researched this phenomenon and found it to be a common occurrence among people who lost a loved one, especially a child—or two in my case. Time became muddled in my brain. Years passed, yet events seemed like they had happened yesterday.

Funny Story

Four years since that fateful day in March, Kim and I once again went to a musical to celebrate Jeffrey's Angelversary. This story started when Melinda visited Kim and me in California earlier in the year. While reading the newspaper, Melinda noticed a musical called *Annie Get Your Gun*, and she asked if we could attend. Of course, Kim jumped right on the idea and bought three tickets for the show at $125 each.

Melinda's health conditions affected the tickets Kim bought, as one of the seats needed to be for a person in a wheelchair. This meant there would be no actual seat for one of us to sit in. We would take a wheelchair to the theater for Melinda.

Unlike when Bud and Jeffrey died, when Kim and I had tickets for this theater shows, nobody died. However, before Melinda could attend the show, she went into congestive heart failure. The ambulance took Melinda to our local hospital, and due to the severity of her condition and her insurance coverage, they flew her home to Fallen Meadow, Utah, in an air ambulance to a hospital covered under her insurance.

That left Kim and me with three expensive tickets to another show. Of course, wasted tickets were not in our vocabulary, so we asked our neighbor, Betty, if she wanted to attend the musical with us.

"Yes, sure," Betty said.

Since the theater was all the way in Poison Oak, California, Kim made reservations: dinner for five. Of course, only three people showed up. Restaurants did not love us, but they always treated us well when they heard the story about the loss of my sons. Besides, we tipped well.

The three of us went into Rodeo, California, and had dinner at this cool Mexican restaurant where the waiters sang the menu; there were no paper menus. Interestingly, with every course of food, they served little shots of an alcoholic drink to cleanse the pallet. Since I did not drink, Kim drank hers and then mine.

The food was unbelievably delicious. We all enjoyed our meals; Kim and Betty enjoyed the alcohol. That wheelchair turned out to be a good thing when Kim had too many drinks. Betty and I had to pour Kim into the wheelchair and roll her into the theater.

Kim has always been a one-drink girl, so it became a comedy just watching her. She even fell asleep during the show. For years, we had the best time sharing this story.

Bud and Jeffrey were surely laughing their asses off up in heaven at the whole situation. I was sure they enjoyed watching us eat, drink, and have a fun time in Jeffrey's honor. He would have loved it and been happy to join us. If only it had been in person.

The Next Day Came.

43

GRADUATION WAS IN SIGHT

The last quarter started on April 2, 2005 and ended on June 7, 2005. Look at that—no 26 or 28 in the scenario. Spring was in the air, and the weather was warm, which was my favorite time. I only had two academic classes this quarter, Community Policing, and the Juvenile Justice System.

Yes, Julie and I took swimming, golf, and weight training again. This activity started to help my physical and mental health. It would take years to recover from the damage done to my body with all the stress these losses had put upon it, along with the hundred pounds of weight gain.

Reminiscing

With graduation pending, my mind wandered back to that scared person I had been four years earlier when Bud called me on The Promise and pushed me to go to Surf City College and pick up a college course catalog. Kim and I still laughed occasionally at my reluctance to take a class because I did not know anything about that subject.

Looking back on that day, Bud saved his mother's life. I doubt that I would have stayed alive without The Promise to fulfill to my two sons. I still smile at how proud of himself he was when he called me on my word. I wondered if he believed that his challenge would force me to move forward to honor my promises. Did Bud ever realize the impact his words had on my life? I always kept my promises, so his one act resulted in me staying alive for the four years it took to earn this degree.

With the challenge of fulfilling The Promise, Bud and I began attending college together. These classes became the sole purpose in my life. With no other reason in my mind to continue living, college became a tool for me, and even for Bud, to somehow survive the loss of Jeffrey.

Unfortunately, college did not turn out the way it started. With the senseless loss of Bud, Kim had to step up once again. She had to take over and get me through another horrible time of loss in my life.

Believe me, there were hundreds of days that it truly would have been easier to end my life and go be with my sons. Somehow, Bud had known, and Kim continued to believe, that I had an important reason to stay on Earth. They knew there were things I was meant to do with my life. So, The Promise became a mechanism that Kim used regularly to keep me moving forward and alive.

Just the Facts, Ma'am

In hindsight, the odds stacked against me from day one of my four-year college attendance.

- I was forty-one years old.
- I started college five months after Jack Ashe murdered my son, Jeffrey, in a robbery.
- I suffered a career ending injury one month before I started college, resulting in a permanent disability to my arm and unemployment.

- Bud, two years later, went to war in Operation Iraqi Freedom and died in a homicide at the hands of undocumented immigrants.
- Three other important people in my life died of natural causes.

The physical and mental effects of the events triggered PTSD, complicated grief, anxiety, and deep depression, which kept me under enormous stress. I also dealt with incredible physical pain from the workman's comp injury. This frozen shoulder involved three surgeries to tear scar tissue. Considering it is an enormous accomplishment to survive any of these individual events in life, I threw in the added pressure of earning a four-year college degree.

These certainly were not the easiest years of college considering all that I went through, but I can honestly say that going to college saved my life. Fulfilling The Promise to my sons became my life's focus, which kept me alive for four years. In reflection, I now know that there was another powerful force in my life. Kim was the force that kept me moving forward.

People often asked me how I fulfilled The Promise to Bud and Jeffrey. It was simple, but complicated, simultaneously. Every day, I did six things whether I wanted to or not—thanks to Kim.

Kim's Rules

1. I got up. (Kim woke me up)
2. I ate food. (Kim fed me—diabetic thing)
3. I dressed. (Kim vetoed any nudity on campus)
4. I did my homework. (Kim would not do homework, as she already had an MBA)
5. I drove myself to campus. (Kim would not chauffeur me)
6. I repeated this process.

Truthfully, there were days these six things took absolutely every freaking thing that I had, just to get through this list.

In addition to Kim's Rules, there were two synchronous things: 1

stayed sober, and I did not kill myself. These were incredibly important items in the scope of things. There were days that it would have just been so easy to drive myself over that *big cliff* at a high rate of speed from the top of the mountain. Days also existed when the numbness by copious amounts of alcohol might have been a gift.

All these things combined for me to fulfill The Promise. However, I would never have made it through any of this without the love, support, and help of Kim. She was a saint and earned her spot in heaven, tenfold. She has been a blessing in my life, certainly a gift from the Universe. I say every day in my mind: *Thanks be to Kim.*

Forest Fires

The only classes I missed during the four years at California State University were because we lived in a national forest that an arsonist set on fire. Kim and I had to evacuate our home and the mountain with no idea when, or what, we would return to. There was no way to get to the university through the fire.

The roads to and from the mountain closed for days; the interstate highway even closed as the fire crossed over it. Kim and I could not get back to our home on the mountain for ten days. Thankfully, we had a home to come home to. The house on the road below ours burned to the ground.

I finished the quarter with an A in all my classes. As a bonus, through my study group, I managed to help other young students along the way. My sense of responsibility, integrity, and dysfunctional personality forced me to fulfill The Promise to my sons.

I persisted with college, earning a bachelor's degree, sadly without Bud. I stood tall, albeit a whole lot heavier and less healthy, but I somehow managed to complete the task that Bud wished for me. Never in my wildest dreams did I honestly believe I would achieve this goal. It was unbelievable; indeed, I earned a Bachelor of Arts degree. I hoped Bud and Jeffrey watched from heaven!

Graduation Day Arrived

On June 18, 2005, I received a Bachelor of Arts with highest honors at California State University.

Kathy and Kim—Graduation 2005

Before the tragic deaths of Jeffrey and Bud, earning a college degree of any kind had not entered my mind. The Promise had been to convince Bud and Jeffrey that with an education, they could go anywhere and do anything. Originally, it had nothing to do with me.

I fulfilled The Promise made to my two sons through all the tragedies and turmoil of the four-year journey to my bachelor's degree. I wore four ribbons around my neck signifying the honors I earned by receiving all letter A grades (4.0 GPA). To me, they were

Bud and Jeffrey's honors, as they had always believed in me through thick and thin. I was proud of myself and the fact that I persevered and survived long enough to earn this degree, for them.

My story should convey determination, hope, and inspiration to anyone who has struggled with a loss, especially the loss of a loved one or even worse, a child, or two. By honoring my two sons' memories, I proved to the other students that they, too, could earn a degree through any adversity.

There are No More Excuses for Failure; Rise and Make It Happen.

Quite a group of friends and *new* family showed up for my graduation. Not surprising to me, neither of my parents, nor any family members showed up for this graduation. Gleefully, as I crossed the stage with my friends Julie and Erin to receive my diploma, I heard people cheering in the auditorium. Unfortunately, Bud and Jeffrey were not there in person, but I knew in my heart they were there in spirit. I could feel their love surrounding me.

I imagined Bud and Jeffrey walking across that stage, receiving my diploma with me. I felt them both give me a high-five. In my heart, I trusted that Bud and Jeffrey watched over me every single day. This day could only have been more extraordinary if Bud and I could have finished this college degree together, graduating as mother and son. How cool that would have been.

Kim arranged a wonderful party at the Oaks Restaurant, up on the mountain. It was good to have people who loved and supported me. I received a dozen presents. Kim even ordered a beautiful cake with candles.

Jerry and Geri had become adopted parents to me. Peggy and Emilie stayed our friends and still are today. Marisol became a daughter to me, a part of our family.

Kim, Kathy, and Marisol

Occasionally, I still hear from Chad, mostly when he finds a photo of Bud or Jeffrey. These people were there to celebrate the impossible accomplishment I had achieved.

All these people hung on to me and held me up when I had no strength or will to continue. First, they were happy I was still on this Earth. Second, these people were proud of me for sticking to The Promise and earning a Bachelor of Arts degree for Bud and Jeffrey. I earned this degree in their honor; Bud and Jeffrey knew I earned this degree for them! It was a huge day indeed.

Reality Hit Me

The reality of the day was there were no emotions left in me. I should have been excited and happy for achieving this accomplishment, considering all the hell I had survived. Yet, inside, I felt nothing. There was just an emptiness in me. I was flatlined.

The Promise to those two little boys kept me going for so long. I worked so hard to complete this goal. I survived such pain, sorrow, and loss. I knew in my heart that Bud was enormously proud of me; he was my biggest supporter. He believed in me when I could not believe in myself. I knew Jeffrey was proud of me, too.

"Mom, you are the smartest person I have ever known," Jeffrey always said.

If there was a heaven above, Bud and Jeffrey were there together, smiling down on me. I just wished they could have shared this accomplishment with me. My life would never be the same without them. For however long I remained on Earth, I would dedicate my life to honoring them. I vowed to keep their memories alive and make them proud every single day.

My goal was to make sure that their lives stood for more than the violence and anger that ended their lives. I wanted their lives to stand for good, powerful, and wonderful things. Their legacies and stories would help thousands of people throughout the world to heal and learn to love one another. I tended to gild the lily when it came to Bud and Jeffrey, but this love was a mother's prerogative.

The Next Day Came. They just kept coming.

44

THE PROMISE FULFILLED

Kim and I bought the plaque pictured below and placed it at California State University outside the Criminal Justice Department. The plaque signified the fulfillment of The Promise along with my undying love for Bud and Jeffrey. Even though I achieved a college education, I had not, nor could I even look for, that easier job this degree was supposed to afford me; I was not ready yet.

Honorary Plaque for Bud and Jeffrey

Oddly enough, receiving this Bachelor of Arts degree left me in another vacuum of emptiness. When I awoke the next morning, I realized I had no idea what I was supposed to do. What was my next step? The Promise had filled the last four years with a razor-sharp purpose. What was I supposed to do with the rest of my life? How would I ever fill the desolation?

New Directions

There was no way I was ready to go back to work as I could not manage a job or even function for eight hours a day. Certainly, I was not able to do anything for five consecutive days a week. Hell, there were still days I was not able to function at all.

Sure, I had started to move forward and heal, but there was grief I had not even begun to walk through or deal with. I still had nowhere I wanted to go and nothing I wanted to do. With the fulfillment of The Promise, it was time to dig deep and find a new purpose in my life.

Big Suggestions

On my last day down at the university, Dr. Primrose called me into her office for a conversation.

"What are you going to do now, Kathy?" She asked.

I looked at her, dumbfounded. "I have no idea, Dr. P."

"Well, we think you should apply for the Master of Arts program here at the university."

"I woke up this morning in a panic. I spent the last four years focused on The Promise I made to Bud and Jeffrey. Somedays, just showing up took everything I had. I'm a bit surprised that I earned this degree."

"Kathy, you did, and with top honors. I know your sons and Kim are immensely proud of you."

"Now, I really have no idea what to do," I said.

"Kathy, you're an excellent student. You mentored dozens of students, helping them succeed. You are such a kind and caring person. The world certainly dealt you a crappy hand of cards. Think about it—since school helped you this far in the healing process, you might need more time to figure it all out."

"I'll ask Kim and see what she thinks."

"Okay, let me know. We'd be happy to have you in the program."

Dr. Primrose became a good friend to me the first day I walked into the Criminal Justice Department. We moved far beyond the teacher-student relationship; we became friends. I appreciated her thoughts and ideas about me. I knew she genuinely cared about my future. We are still friends, even today.

"Thanks for all your help, Dr. P. I hope Bud and Jeffrey are proud of me."

"Oh, I know they are. How could they not be proud? You did it for them," Dr Primrose said.

Kim, of course, did not hesitate for a second. She encouraged me to continue in my education. *What would my world be without Kim?* I applied for the program, and they accepted me into the Fall, 2005 Master of Arts program for Criminal Justice.

Another Summer with Nothing to Do

I had no idea what I signed up for with the next program, but the previous four years taught me that I could learn anything. Going to college had kept me alive for four years. I was smart enough to know that if something worked, I should keep doing that thing until I figured out why the Universe had left me on Earth without Bud and Jeffrey.

People often asked me how I survived the loss of my two sons— my only children—so I reflected on this whole going-to-college experience in my life. I never verified this, but I believed that the American Psychiatric Association would not have recommended four years of college as a treatment for a mother who violently lost both her children. However, they might want to review my case, after diagnoses of chronic conditions, not limited to, complicated grief, PTSD, migraines, high blood pressure, and other stress-related issues. I do know that four years of college saved my life and allowed me the time to start healing.

I wondered if Julie and or Erin had applied for the program; I should have asked them. It would be nice to spend more time with them. I hoped Bud and Jeffrey were paying attention, as we had never discussed an advanced degree when I originally made The Promise. *Do Bud and Jeffrey receive updates in heaven?*

Sadly, the program would not begin until September 22. That left me the whole summer to contemplate and worry about this new adventure, a distraction from death. I still had thousands of questions in my head. I hoped that someday I would find answers or a peace with which I could live. If I did not find this at college, I prayed that I would find resolutions somewhere in my life.

In my mind, I pictured a light bulb or little speech bubble coming on above my head, like in the comics or cartoons, when I received those answers. Unfortunately, it had not happened yet. I held onto the hope that I would find those answers in this Master of Arts program.

The Next Day Came. Each morning I opened my eyes, and I was still alive.

THE PROMISE FULFILLED
SUPPORT STEP

As I learned more about coping with my own loss, many recommended journaling and coloring as powerful tools. Portions of this book may trigger intense feelings, good or bad, while reading these stories. If this happens, write those feeling down. To help you with this, I created **The Next Day Came Trilogy Thoughts and Emotions Activity Book**, which provides space to write and color when you need to take the time to process your thoughts. Along with this resource, I am also providing you with the Limitless Resilience Checklist and How to Build Emotional Resilience Video to help you become more resilient in the face of extreme adversity.

These resources can be found here:
www.LimitlessResilienceKit.com
Or scan this QR Code:

Honor Your Losses, Love, and Live Life Limitlessly,

KD Wagner

CONCLUSION

There will never be closure for me around the loss of my children. My belief stays firm that there is no such thing as closure or justice; it is just an illusion for the victim's loved ones left behind. They caught the perpetrator in Jeffrey's murder, sentenced him to time in prison, but absolutely nothing changed for me—my son was still dead. I experienced no less pain; I found absolutely zero peace.

My beliefs about the lack of justice and closure in our criminal justice system continuously verified itself in my life. Yes, Jack Ashe spent eleven years and ten months in prison, but everyday life continued for him and his family. Jack Ashe had friends, watched television, read books, ate three meals a day, or as the Marine Corps Sergeant told us: *Shit, shower, and shave* happened; the day-to-day intricacies of life went on.

In fact, Jack Ashe was a tattoo artist. At one of the parole hearings, Kim and I learned that Jack Ashe made his spending money for commissary items in prison by doing tattoos for inmates. He enjoyed all the things he did before and after he murdered Jeffrey. In fact, he was quite popular. Jeffrey's life ceased to exist; Jeffrey was still dead.

Adding insult to injury, the California Highway Patrol botched

Conclusion

the investigation into the felony hit-and-run that caused Bud's death. To the date of this publishing, nobody from the California Highway Patrol has found the perpetrators in Bud's homicide, nor have I ever heard another word from them.

The vehicle, with three undocumented immigrants, crashed into Bud on his Harley-Davidson motorcycle. They knew they were in trouble, raced off, and left Bud to die. They, along with their families and whoever else lived in that house, packed up and moved out while the police were not watching the house. They nonchalantly drove off to another state or left the United States of America altogether. Either way, they outsmarted the California Highway Patrol and escaped law enforcement.

Since August 28, 2003, none of the three undocumented immigrants have committed any further crimes for which the police made an arrest, nor have they applied for a job that needed fingerprints taken. If any of these would have happened, their fingerprints would have popped in the national database about Bud's death.

In my mind, these three men are living their lives, just like the monster that murdered Jeffrey. They breathe air, enjoy friendships, watch television, and enjoy food every day. Unfortunately, Bud is still dead.

Sadly, on the evening news, Kim and I often see the mother and or father of a murdered child crying for their lost child. They demand closure and justice. Sadly, Kim and I look at each other and say the same thing every time: *There goes another family that will never be the same; ruined, they will never find closure or justice.* When will this stop? What *important dead child* would give reason for these gun nuts to change the laws? When will people stop shooting other people?

In my heart, closure settled in my life as a fantasy. My advice for other parents or people who suffer a loss of any kind, would be to not waste their time wishing and hoping for justice or closure. The best thing I have found that helped me was the hope for karma to do the work that law enforcement could not.

If you are suffering a loss, try to find a new purpose and passion for life. Move forward from the loss one step at a time. I learned to focus on becoming a better person, helping others, or giving back. I did the things Bud and Jeffrey could no longer do. By trying to live my life to its fullest, I honor Bud and Jeffrey, making them proud in everything I do.

Looking outside of myself allows me to be more kind and helpful to others. Doing these things give me a sense of peace. I love these two boys so much. There are days when all I can do is to continue breathing. These boys were my everything. I move heaven and Earth to make them proud of me every day. I learned how to love, respect, and treat people in a healthy way from Bud and Jeffrey. They truly were a gift to me, a blessing from the Universe.

Another point revealed to me through the loss of my two sons was that family, friends, and people in general do not know how to deal with or express feelings about loss. They have no idea what to say or how to comfort the person who has experienced a loss of any kind, much less the ultimate loss of a child, or two.

When drowning in my deepest despair and depression, people disappeared from my side. Just when I needed family and friends the most, they vanished. They appeared clueless on how to be supportive. I lost so much more than just my sons. I lost my entire family, dozens of friends, and my career.

People always point out: *Well, you had Kim*. I can guarantee that I would not be here today writing this story if not for Kim, but even she had no idea how alone I truly felt because there were no words to describe my losses.

My parents, my siblings, their families, or my extended family do not talk to me. They rarely contact me in any way, shape, or form. If it were not for Facebook, I would not know they or their families were alive. I sometimes see what they are up to as they post about weddings, graduations, children, and grandchildren.

My family absolutely refuses to talk about Jeffrey or Bud. In fact, they never mention their names. If I happen to mention them, the subject quickly changes. Somehow, they seem to believe that death might be contagious and that by mentioning my children, or the death of my children, their children might also die or another absurdity.

Conclusion

My parents, siblings, and friends never recognize Bud or Jeffrey's birthdays or Angelversary days, other than an occasionally like on Facebook when I post about the day. They do not call me on holidays, nor do they ask if I am okay. They never share any memories they have about Jeffrey or Bud. I know they have stories, memories, and even photos of my children, as their children grew up with Bud and Jeffrey. It truly is the weirdest thing, beyond painful and hurtful to me.

These phenomena opened my eyes to how people believe they should not talk about loss or death. Trust me, it will never hurt my feelings if you mention my sons or ask me questions about them. In fact, my heart will swell with pride if you remember Bud and Jeffrey. People should realize they inflict even more pain by not talking about loss, especially the loss of a child or loved one. I have not forgotten Bud or Jeffrey for one single second of my life.

The most powerful thing that helped to move Bud and me forward when Jeffrey died was the fact that Jeffrey would forever be alive in our hearts. We talked about Jeffrey, as we tried to figure out what happened to him and why. We shared funny stories about Jeffrey, and together we laughed and cried. Bud and I made a pact that we would strive to be the people Jeffrey believed us to be. This became the most important focus for Bud and me.

When Bud died, I continued that same pattern. I used the exact things that kept me alive after Jeffrey died. I believed in my heart that if these things helped me to survive the loss of Jeffrey, then these same things would be the only hope I had to survive the loss of Bud.

Even today, I particularly enjoy telling stories and sharing the love I have for Bud and Jeffrey whenever and wherever the opportunity avails itself. I not only wrote this trilogy preserving their legacies, but I also speak at events sharing their stories and lives. My humor, albeit dark sarcastic humor, continues to help me even though it sometimes shocks people when snide things come out of my mouth. Other times, people do not even catch the humor in what I say, but I get a kick out of it, and I know Bud and Jeffrey would too.

Conclusion

"Kathy, you are your own best friend, I swear," Kim says.

I look up to the sky every day and talk to my children. I believe in my heart they watch over me, protect me, and guide me in all I do. I will forever miss them. Bud and Jeffrey live on through their stories and are forever in my heart. The tragedy of their loss forever changed my life. However, through the healing process, I became a more caring and loving person, in their honor.

The Next Day Came. Every day, I wake up to opportunity, filled with a new and powerful purpose.

Our story continues in book three of the trilogy: *Kathy: How to Live Life After Loss*.

AFTERWORD

REMEMBRANCES FROM BUD'S TIME IN THE NAVY

Bud's friend Don and I reconnected on Facebook when it came about. I enjoyed getting to know Don through Bud, and we remained friends after his death. Still today, Don visits with Kim and me when he comes to Florida. It warms my heart every time I hear from him, as it somehow provides me with just a little piece of Bud.

I know Don carried the memories of Bud deep in his heart, just like me. Their friendship was extremely important to him, and his loss was deep. Don credited Bud with turning his life and his Navy career around. Whenever we talk, I always tell Don: "Bud is so proud of you, Don, and all you've accomplished. He's watching."

Don retired as a Chief Petty Officer from the U.S. Navy Reserves. Watching Don was a double-edged sword as it reminded me of all the dreams Bud could have accomplished had he lived.

Heartwarming Story

On November 11, 2017, I noticed Don chatting on Facebook about Veteran's Day. I typed *Hello,* and we shared two or three exchanges of conversation.

Afterword

The men who were chatting with Don did not know who I was, but Don must have messaged them and told them I was Bud's mom. Next thing I knew, I received the following messages.

> *Kathy, I had the distinct pleasure of serving with your son and would like you to know that not only was he an exceptional Sailor, but he was also an even better friend. Although we were peers in the Navy, I saw him more as a mentor.*
>
> *Bud shared his professional knowledge with all that asked, and he personally invested himself for his fellow shipmates, all the while he was able to make each day more enjoyable, even when we were working long hours on tedious taskings.*
>
> *I was saddened to learn of his passing and have shared his memories quite often over the rest of my career with my own Sailors. ~ Paul*

The second message was just as sweet to me. Obviously, fourteen years later, these men had not forgotten Bud or how he altered their lives.

> *Paul, my respect for you just skyrocketed! That's awesome brother!*
>
> *Kathy – Had no idea you were Bud's mother. You raised one outstanding son! He helped change the paradigm, at least in our shop, if not the squadron.*
>
> *Before Bud, most reservists would sit around talking, reading the paper, or staying away from the shop altogether. Saying they had an appointment, and they would be gone all day. Lol!*
>
> *Not Bud! He asked to be put on jobs and inquired what it took to get qualified to be an inspector. Then he did whatever it took to do it.*
>
> *The active side always complained instead of helping the reservists learn. After they realized that these guys can be an asset, every one of*

Afterword

the work center 120 reservists were on the way to getting fully qualified, thanks to Bud.

We also talked a lot about his bike. Lol.
B E Z and Happy Veterans Day, Bud's Mom! ~ Mike

It was heartwarming for me to hear such kind thoughts and memories of Bud after all these years—how he had changed grown men's lives. It still breaks my heart to imagine where Bud would have gone if only people had not senselessly killed him that day on his way to go back to war. What a waste of a good life.

ABOUT GOLD STAR MOTHERS, INC.

When a person joins the United States Military, national guard, or reserves, their mother, stepmother, grandmother, foster mother, or female legal guardian becomes a Blue Star Mother. Unfortunately, the Department of Defense does not give any information about the Blue Star Mothers of America, Inc. to the person joining the military, nor do they send information to their immediate family members.

The Blue Star Mothers of America, Inc., have chapters throughout the U.S. As a non-partisan, non-sectarian, non-discriminatory, and non-profit organization, they support their members and people who currently serve in the U.S. Armed Forces, along with veterans and Gold Star Families. To find a Blue Star Mother group near you, visit their national website: www.bluestarmothers.org

There are unlimited benefits of being a Blue Star Mother. They support deployed military personnel, for example, by sending boxes full of snacks, necessities, cards, and letters. One of my favorite projects is *Red Shirt Friday*. The premise is that every Friday people wear a red shirt to remind others in the community to remember the deployed men and women throughout the world, as *Freedom is Not Free!*

If you have a child serving in the U.S. Military or know someone who does, they are eligible to join the Blue Star Mothers of America, Inc. Please contact them at the link above.

If the unimaginable happens and a child dies while serving on active-duty in the United States Military, then their mother becomes a Gold Star Mother. This group is a 501(c)(3) national non-profit

Veterans Service Organization (VSO), founded in 1928, incorporated, and chartered by the United States Congress. The President of the United States declares a proclamation the last Sunday in September every year to honor Gold Star Mother/Family Sunday.

Most all volunteerism happens through local chapters throughout the country. To find a chapter near you, go to their website: www.goldstarmoms.com. Gold Star Mothers serve veterans and their families by sitting on the Advisory Board of the Veterans Administration Voluntary Services (VAVS). Each year, Gold Star Mothers donate thousands of dollars, hours, and donations to Veteran Hospitals, Veteran Centers, and National Cemeteries.

On a personal note, I never knew I was a Blue Star Mother until I became a Gold Star Mother, in 2017, thirteen years after the loss of Bud. At that time, I joined a chapter near my home in Florida. I currently serve as the president of the Gulf Coast Chapter.

Looking back at my experiences, I realize that with no further contact from the military and the non-contact with other Gold Star Mothers, my grief, loss, and depression continued for an unnecessary extenuated time. As I shared in my book, at the time the letter from the Department of Defense arrived at my home, I had no comprehension of what being a Gold Star Mother meant.

I wish the American Gold Star Mothers Inc. group would have contacted me right away after the loss of my son Bud. I believe in my heart that having this group present earlier in my grief would have sped up my process in healing by having friendships of like-minded mothers in loss.

Being a part of the American Gold Star Mothers, Inc. has helped me heal and move forward by giving back and helping others. Surrounded by other mothers who have lost a child while serving in the military has brought me comradeship. Through these experiences, I have learned to share my story with people around the world.

I share this information to ensure that every mother knows that the Blue Star Mothers exist and are there to support and help them. Additionally, no mother who lost a child serving in the United States

Military should ever remain in the dark about the Gold Star Mothers and the good they do for each other and the community.

The importance of knowing that somebody understands your loss and pain is a wonderful gift through the process of grief and recovery. Together, we learn to live life again with a new purpose and passion to honor our children.

If you or anyone you know lost a child on active-duty while serving in the United State military, please have them contact the American Gold Star Mothers, Inc. at the link above to find a local chapter in their area.

TAKE THE NEXT STEPS

Now that you have read *Bud*, it is time to learn the rest of the story. Read all three books in *The Next Day Came Trilogy*.

Book One: *Jeffrey: The Injustice of Murder*
Book Two: *Bud: Homicide Turns a Blue Star Gold*
Book Three: *Kathy: Survive Tragic Losses with Limitless Resilience*

Find the eBook / Paperback / Hardcover / Audiobook at www.limitlessresilience.com

In Book One, *JEFFREY,* and Book Two, *BUD, you found* hope, and tips for moving forward. If you – or someone you know – would benefit from additional support on this journey, below are some of the ways you can continue to work with me. You can find more information at www.limitlessresilience.com:

- *Limitless Resilience Kit*
- *Limitless Resilience Mini Course - video course*
- *6-week Limitless Resilience Course (self-paced videos and weekly group coaching)*
- *Limitless Resilience Community Mastermind*
- *Limitless Resilience Private Coaching*
- *Limitless Resilience VIP - spend the day with Kathy*
- *Limitless Resilience Podcast*
- *Customized Speaking engagement*

I encourage everyone to:

- Take photos of your children and loved ones. You will never have enough!
- A donation of all book proceeds will go to The American Gold Star Mothers, Inc., Gulf Coast Chapter, supporting veterans and their families.
- Donate to The American Gold Star Mothers, Inc., Gulf Coast Chapter to support those who have lost a child: www.agsmgulfcoast.org
- Donate to Operation Support Our Troops—America

1807 S. Washington St., Suite 110, #359
Naperville, IL 60565
www.osotamerica.org

- Donate to Island Dolphin Care

150 Lorelane Place
Key Largo, Fl 33037
www.islanddolphincare.org/get-involved/donate/

ABOUT THE AUTHOR

Dr. KD Wagner is a multiple #1 Bestselling Author, Eippy Award winner, international speaker, and former law enforcement officer. Featured in her Trilogy, *The Next Day Came*, revealing her journey through depression, addiction, planned suicide, and survival after the unimaginable violent loss of her two sons in separate homicides, two years apart, at 18 and 24.

Dr. Wagner is a Gold Star Mother—whose son died while serving in the U.S. Navy—during Operation Iraqi Freedom. KD currently serves as the President of the Florida Gulf Coast Chapter within the American Gold Star Mothers, Inc. Her volunteerism supports veterans, their families, and the community, speaking on and promoting awareness, to raise contributions for their organizations.

Dr. Wagner has helped thousands of people on the topics of loss, survival, and thriving. She speaks on the loss of a child, gun violence, grief, addiction, learning disabilities, and how to move forward with strength and courage. With profound empathy, KD Wagner inspires people to find the purpose in their pain, find a new purpose, and find life after loss.

Dr. Wagner earned both a bachelor's and master's degree from California State University with the highest honors to fulfill The Promise, and in Honor of her two sons.

Dr. Wagner has appeared on NBC, ABC, CBS, FOX, and stages with Joint Chief of Staff General Mark Milley, Governor/Senator Rick Scott, country singer Rockie Lynne, Bob Circosta of the Home Shopping Network, Nancy Matthews-*Women's Prosperity Network,* 'Dr. Lydie Louis Esq.-The Money & Law Doctor, Gary Coxe-Business and Life Strategist, J.T. Foxx, Orly Amor-Health and Wellness Network, and Shannon Procise (Gronich)-*Business Acceleration Network.*

Dr. Wagner took the Leap of Faith and *jumped* at 14,000 feet with the U.S. Army Elite Parachute Team—The Golden Knights and *bent steel* with Bert Oliva. In her journey, she has discovered ways to not only survive, but to thrive under extreme circumstances with Limitless resilience.

Dr. Wagner's mission is to honor her sons in everything she does and be the person they believe her to be, living every day to make them proud in all she says and does.

Dr. KD Wagner currently resides in Florida with her spouse, Kim, and their standard poodles.